Geckos: *The Animal Answer Guide*

Geckos

The Animal Answer Guide

Aaron M. Bauer

The Johns Hopkins University Press   Baltimore

© 2013 The Johns Hopkins University Press
All rights reserved. Published 2013
Printed in the United States of America on acid-free paper
9 8 7 6 5 4 3 2 1

The Johns Hopkins University Press
2715 North Charles Street
Baltimore, Maryland 21218-4363
www.press.jhu.edu

Library of Congress Cataloging-in-Publication Data

Bauer, Aaron M.
  Geckos : the animal answer guide / Aaron M. Bauer.
    p.   cm.
  Includes bibliographical references and index.
  ISBN 978-1-4214-0852-1 (hdbk. : alk. paper) — ISBN 978-1-4214-0853-8
(pbk. : alk. paper) — ISBN 978-1-4214-0925-2 (elecronic) —
ISBN 1-4214-0852-X (hdbk. : alk. paper) — ISBN 1-4214-0853-8
(pbk. : alk. paper) — ISBN 1-4214-0925-9 (electronic)
  1. Geckos. I. Title.
  QL666.L245B375 2013
  597.95'2—dc23   2012027048

A catalog record for this book is available from the British Library.

*Special discounts are available for bulk purchases of this book. For more information,
please contact Special Sales at 410-516-6936 or specialsales@press.jhu.edu.*

The Johns Hopkins University Press uses environmentally friendly book
materials, including recycled text paper that is composed of at least 30 percent
post-consumer waste, whenever possible.

The Gecko lying on its stone
Is always very much alone,
Nor is the reason hard to trace
By those who've seen its form and face
It's hard to realise a mite
Can be so venomous a sight,
Or in its little frame compress
Such concentrated ugliness . . .
Yet underneath its skin, we're told,
There beats a heart of purest gold.
Its children do not know neglect;
It treats its mother with respect . . .
Its aspect is its sole defence
Against the world's malevolence.

From "The Gecko" by LEON GELLERT
Beastly Australians, Hodder and Stroughton,
London, 1964

Contents

Acknowledgments

This book had its birth in the encouragement I received from my mother to follow my herpetological interests. It was made possible by the training I received from my teachers and academic mentors, particularly Marvalee H. Wake, who waded through the 869 pages of gekkotan minutia that was my doctoral dissertation. It was informed by what I learned from many of the "gecko gurus" whose works I read and with whom I corresponded and conversed, among them Garth Underwood, Nikolai Szczerbak, Yehudah Werner, Wulf Haacke, Don Broadley, Max King, Bob Bustard, Arnold Kluge, Paul Maderson, Hal Cogger, and Glen Storr. It also benefitted from the wisdom imparted by my many field companions and coauthors. From this nearly endless list I will mention by name only Ross Sadlier, Bill Branch, Johan Marais, Indraneil Das, Trip Lamb, David Good, Todd Jackman, le Fras Mouton, and Tony Whitaker, with whom I have had particularly long and enjoyable friendships (to all my other friends my thanks, though implicit, are just as heartfelt).

Since becoming a graybeard myself, my postdocs and students have been a source of inspiration for me and have kept me on my toes. While they have all done their part, I will single out the most "geckoey" among them: Eli Greenbaum, Juan Daza, Matt Heinicke, Tony Gamble, Stu Nielsen, Heather Heinz, Phil Skipwith, Jennifer Bourque-Wright, and Scott Travers.

Ultimately, this book exists because Vince Burke at the Johns Hopkins University Press convinced me it was worth doing, because Chad Peeling convinced me that there was an interested audience of closet gecko-lovers, and because my wife, Monica (and our Maltese, Bertie), gave me the inspiration and the time to actually write it. The book could not have been completed without the kindness of Juan Daza and Norman Dollahon, who helped prepare some of the images and Tony Gamble, Johan Marais, Bill Branch, L. Lee Grismer, Tony Whitaker, Brad Maryan, Jon Boone, Randy Babb, Joshua Snyder, Ashok Captain, Laurent Chirio, Miguel Vences, Mark O'Shea, Mike Bartlett, Dennis Hansen, Anslem de Silva, Trip Lamb, Mirko Barts, Bruce Thomas, Jean-François Trape, Jeff Wright, and Wolfgang Weitschat, whose excellent photos illustrate the text.

I reserve a final thank-you for Tony Russell, who has been my postdoctoral advisor, my best man, my closest colleague, and my oldest friend and who remains my greatest role model. Happy sixty-fifth birthday!

Introduction

When I was growing upon New York's Long Island in the 1960s and 1970s, "gecko" was hardly a household word, and even budding herpetologists had little familiarity with these exotic lizards. However, the grass is always greener on the other side. I, having snakes, turtles, frogs, and salamanders on my doorstep, of course, became fascinated with lizards, which (except for one introduced species) were absent locally. My first exposure to lizards, and to geckos, came via Roger Conant's *Field Guide to the Reptiles and Amphibians of the United States and Canada East of the 100th Meridian* that I received at the age of 6. Plate 14—which showed the Reef Gecko, the Ashy Gecko, the Yellow-headed Gecko, the Banded Gecko, and the Mediterranean Gecko—was engrained in my mind. I looked forward to a future in which I would encounter geckos in the wild. No one ever discouraged me from herpetological pursuits, so I made studying reptiles, and particularly lizards, a career goal (in junior high school shop class I printed up business cards that said, "Aaron M. Bauer, Herpetologist"). Eventually, in 1983, as a graduate student at the University of California at Berkeley, I went on my first collecting trips abroad (Mexico, followed by a round-the-world research expedition with fieldwork in Israel, Thailand, Australia, New Caledonia, New Zealand, and Tahiti), and my gecko dreams were realized. For the nearly 30 years since, geckos have been the focus of my research and my primary professional interest.

Much of the interest in geckos comes from people who enjoy keeping and breeding these beautiful animals. In the interest of full disclosure, I am not now and never have been a big fan of keeping reptiles as pets; I have a "black thumb" when it comes to keeping animals alive in captivity. However, I do not begrudge this pleasure to those who find it rewarding, especially as many such people develop a greater appreciation for biodiversity and the natural world through their exposure to these lizards. Indeed, much of what is known about many aspects of geckos, we know from a few well-studied species of captive geckos, so these are disproportionately represented in this book. I have also mentioned some specifics that might be of particular interest to those keeping geckos in captivity. This book, however, is not intended as a guide for keeping or breeding geckos. Those wanting to do so should consult one of many reliable books now available on the subject. Rather, I have written this book for the reader who has a general interest in geckos and would like to understand their diversity and how they live. The questions answered touch on most of the aspects of

Aaron Bauer—shown here holding a South African Velvet Gecko, *Homopholis wahlbergii*—has been studying gecko evolution since 1982.

gecko biology, and I have drawn examples from as many different gecko groups as possible. In a short book like this one, only a glimpse into the world of geckos can be provided. I hope that readers will find both their questions answered and their curiosity piqued and that they will gain a greater appreciation for geckos as fascinating and important players in the evolution and ecology of life on earth.

Geckos: *The Animal Answer Guide*

Chapter 1

Introducing Geckos

What are geckos?

Geckos are a type of lizard. They are generally small-bodied, fully limbed, nocturnal predators that have good vision and vocal capabilities, usually lack movable eyelids, and often have adhesive pads on their toes. Lizards and snakes, along with amphisbaenians (worm lizards), constitute the order Squamata, or scaled reptiles. Squamates are one of several major groups of vertebrates. They are amniote vertebrates, meaning that they belong to the group of vertebrates that has "escaped" from reliance on water for reproduction by evolving an egg that contains an amniotic membrane. Water is kept in but gases, like oxygen, can move freely in and out of the egg. This lets the young to undergo development in even the harshest arid environments.

Other living amniotes include mammals, birds, turtles, and crocodilians. The last two of these are often combined with squamates in the Class Reptilia. Although this group is familiar to almost everyone, it is not a "natural" group in an evolutionary sense. In other words, some reptiles are more closely related to nonreptiles than to other groups of reptiles. Thus, crocodilians are more closely related to birds than they are to the superficially more similar lizards. We know this to be true because there is a great deal of evidence from comparative anatomy and paleontology and from genetic data. Lizards, crocodilians, and another group of reptiles closely related to squamates—tuataras—are similar in overall appearance, because they walk on four legs and have a scale-covered body. They also share physiological similarities such as ectothermy (a reliance on external heat sources to maintain body temperature) and poikilothermy (the inability to physiologically regulate temperature). These similarities confused scientists like Carolus Linnaeus (1707–1778), who originally classified not only crocodilians but even salamanders (a type of amphibian) as lizards. Only in the early nineteenth century were the distinctions between amphibians and reptiles clarified and the great differences among groupings (families) of lizards recognized.

Where does the name "gecko" come from?

The word "gecko" or "gekko" is widely used in English and many other European languages to refer to fully limbed gekkotan lizards. It was in wide use by the mid-eighteenth century and was used by Linnaeus in 1758 when he described the Tokay Gecko as *Lacerta gecko* (now *Gekko gecko*).The earliest European works in natural history used the Greek-derived name *ascalabotes* or the Latin *stellio* (both meaning a spotted lizard) or *lacertus facetanus* (clever lizard) for geckos. "Gecko" did not appear in Western literature until the seventeenth century, when commercial trade with tropical Asia was well established. The first known usage of the word in the Western world was in a book by Bontius (1658) in which he discussed the animals and plants he had encountered in the East Indies (modern Indonesia) some 30 years before. This and other early references suggest that the name was an imitation of the vocalizations made by this lizard. The *Oxford English Dictionary* derives "gecko" from the Malay *gēkoq* (sometimes written as *kēko* or *gago*), itself an onomatopoeic word based on the sounds made by the Tokay Gecko (*tāké*, *goké*, and *tōkē* are also Malay words for gecko). An alternative possible origin comes from Sri Lanka. Although the modern Sinhalese (the language of the largest ethnic group in Sri Lanka) words for geckos are *hūna* or *sūna*, Sri Lankan works written between the eleventh and fifteenth centuries record *gēgo* (*gē* = house + *go* = *goya* = lizard = house lizard) as the Elu (the precursor of modern Sinhala) name for gecko. Dutch ships bound for the East Indies in the seventeenth and eighteenth centuries regularly called at Sri Lankan ports en route to Ambon and Java, but by this time *gēgo* was no longer in local use. However, cultural, political, and military connections between Sri Lanka and southeast Asia between the sixth and thirteenth centuries may have provided opportunities for linguistic influences between the two regions. Despite the apparently different meanings of the words, *gēkoq* and *gēgo* may ultimately share a partly common origin.

Why are geckos important?

Geckos are important for several reasons. They are key components in many tropical and subtropical ecosystems, from deserts to rainforests,

The Spotted Dtella (*Gehyra punctata*) is a typical nocturnal gecko, with well-developed limbs, smooth skin, padded toes, and a spectacle—a clear eye covering—instead of movable eyelids. Courtesy of Brad Maryan

where they are both predators and prey. Most geckos feed on insects and other arthropods and contribute to controlling their populations. Because most geckos are nocturnal, they play a particularly important role in regulating populations of night-active insects. Indeed, geckos may be second only to bats in this regard. The magnitude of their influence on food webs can be appreciated when one realizes how numerous geckos may be. Although most geckos are small, some can reach high densities. The world's densest terrestrial vertebrate is the Dwarf Gecko *Sphaerodactylus macrolepis* from the Virgin Islands, which has been measured at densities of 21,367 per acre (52,800 per hectare) and a biomass of about 8.9 pounds per acre (10 kilograms per hectare), which is similar to that of the African elephant, *Loxodonta africana*. Geckos are important prey items in the diet of many different vertebrates and even some invertebrates. In some cultures, they have also been eaten by people.

Throughout much of their range, some geckos are commensal with humans. This means that one group—in this case, geckos—benefits, while another—in this case, humans—neither benefits nor is harmed. Although one could argue, by feeding on disease vectors like mosquitoes and food pests like cockroaches, geckos provide a great benefit. Mostly because of this service, geckos are considered to be good luck in many parts of Asia and are of cultural importance. Geckos have also long been believed by practitioners of traditional medicine to have curative powers, and gecko products have been shown to have pharmacoactive properties. Geckos and their amazing climbing abilities have recently served as the inspiration for artificial adhesives and for lightweight climbing robots with applications in a variety of fields from intelligence gathering to surgery.

These illustrations of the Common Wall Gecko (*Tarentola mauritanica*) appeared as *lacertus facetanus* (clever lizard) in a book by Ulisse (Ulysses) Aldrovandi in 1637, one year before the first-known European use of the Asian-derived word "gecko."

Why should people care about geckos?

Geckos are part of the natural heritage of many areas of the world, particularly the biodiverse tropics. Like all components of natural ecosystems, they should be appreciated and their right to exist respected. We should care about them not only because they play critical ecological roles as diverse and abundant predators and prey but because their only interactions with humans are beneficial, controlling insect populations while posing no threat of danger or disease. Both the bold, brilliant colors of the day geckos and the more subtle earth tones of many desert species are aesthetically pleasing and, because geckos are conspicuous in and around human habitations, they are reminders of enjoyable vacations or adventures to visitors to the tropics. In North America and Europe, where native gecko diversity is low and geographically limited, they are among the most common starter pets for children (Leopard Geckos and Crested Geckos) and thus serve as ambassadors of the natural world, helping to establish an appreciation for life's diversity and providing a tangible link to the wild.

Where do geckos live?

Geckos are found on all continents except Antarctica. They are also absent from Arctic areas like Greenland and Iceland and from the northern parts of North America (including all of Canada and most of the United States) and Asia. In temperate North America, native Banded Geckos (*Coleonyx variegatus*) are found only as far north as southern Utah. In temperate Asia, *Alsophylax pipiens* extends as far north as Mongolia. Geckos are most diverse in tropical and subtropical parts of the world, where they occupy all mainland areas and all but a few of the most remote oceanic islands. In the Southern Hemisphere, the Harlequin Gecko (*Tukutuku rakiurae*)

Geckos: The Animal Answer Guide

The Mediterranean Gecko (*Hemidactylus turcicus*) is an important nocturnal predator on insects and is one of many geckos that can coexist with humans. Courtesy of Tony Gamble

is found as far south as Stewart Island (47° S), and *Homonota darwini*, the southernmost gecko in the world, reaches almost to Tierra del Fuego at the southern tip of South America (52° S), but no geckos occur on Tasmania. Four species of geckos survive in Europe, but these are all limited to the south of the continent where they experience a mild Mediterranean climate. Some geckos are good colonizers and have taken advantage of human movements to establish themselves in places where they did not occur naturally. For examples some remote oceanic islands, like the Hawaiian Islands in the Pacific and St. Helena in the Atlantic, were never reached by geckos on their own but have suitable climates to support species that have arrived as stowaways. Parts of the southeastern United States have also been colonized by alien geckos. The Mediterranean Gecko (*Hemidactylus turcicus*) has been especially successful in spreading across much of Florida and the Gulf Coast. There are even some isolated populations as far north as Virginia and Maryland.

The same intolerance of freezing temperatures that limits geckos latitudinally also limits their distribution in elevation. Among the highest living geckos are the Bent-toed Geckos *Cyrtodactylus tibetanus*, *C. medogenesis*, and *C. zhaoermii* from the Himalayas of Tibet and the Atlas Day Gecko *Quedenfeldtia trachyblepharus* from the High Atlas Mountains of Morocco, all of which reach elevations of about 13,000 feet (4,000 m). Geckos are widespread

The Leopard Gecko (*Eublepharis macularius*), a native of arid western Asia, has become one of the most popular of all reptile pets and is bred in huge numbers in captivity. Courtesy of Tony Gamble.

in areas that provide an appropriate thermal environment and some sort of hiding place for them. Many species are terrestrial and may retreat to burrows—either excavated by themselves or taken over from scorpions, spiders, or other small animals—or to protected areas underneath stones or vegetation. Others are scansorial (climbing) and may be arboreal (living in trees), or rupicolous (associated with rocks).

What is the difference between geckos and other lizards?

Geckos differ from other lizards in several ways but because they are so diverse themselves it is not possible to identify any single feature that all geckos—and only geckos—share. Nonetheless, nearly all geckos "look like geckos" and can easily be distinguished from other lizards with a little practice. Gecko skulls are more open in construction and consequently more highly kinetic (mobile) than those of other lizards because they lack two of the bony struts (the supratemporal bar and the postorbital bar) that are present in most non-geckos. They also have a single (rather than paired) frontal bone, one of the main bones of the skull roof. The ventral side of this bone forms a tube enclosing the nerves that carry sensory information from the nose (the bottom of the frontal is not enclosed in most lizards). Compared with most lizards, geckos usually have large heads with rounded snouts, many small teeth, and large eyes. Eye size is probably related to the fact that geckos evolved as nocturnal predators. This night-active evolutionary history also accounts for the absence in geckos of a parietal foramen, an opening in the skull that allows light to reach the parietal eye, a light sensitive organ that is involved in the regulation of daily and seasonal

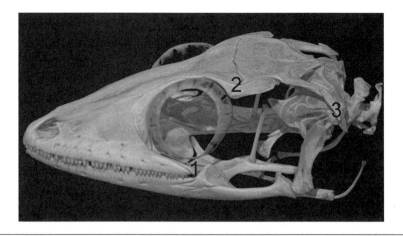

Some of the features that distinguish geckos from other lizards are internal. This x-ray computed tomography (CT) scan of the skull of the Atlantic Forest Gecko (*Phyllopezus lutzae*) of Brazil shows the typical lightly built skull of the group. Most other lizards have a postorbital bar that runs from the jugal bone (1) to the postfrontal bone (2) and a supratemporal bar connecting the postrfrontal to the squamosal bone (3). The loss of these bones in geckos gives their skulls great mobility, which enhances their ability to feed on certain types of prey.

rhythms. Most geckos lack moveable eyelids and instead have a clear scale that covers and protects the eye, known as a "spectacle" or a "brille."

Geckos often appear to be softer skinned than most lizards. This is because their scales lack the reinforcement of bony plates called "osteoderms" that are common in some other groups. In addition, their skins are usually covered by small granular scales that can give the body a velvety appearance. Even if larger pointed or keeled scales called "tubercles" are present, they are usually arranged within a background of finer granules. With only a few exceptions, geckos lack the shiny overlapping scales that are typical of skinks and the crests, casques, dewlaps, and other adornments that are common in iguanian lizards (iguanas, agamids, and chameleons). Sticky feet are perhaps the structures for which geckos are most well-known, however, about 35 percent of gecko species lack toepads, and one whole family of geckos is even limbless. Some other lizards, like anoles (Family Dactyloidae), also possess toepads, but only in climbing geckos are the hindfeet arranged symmetrically (versus having the fifth toe separated from the other four) in order to distribute forces evenly across the adhesive structures.

Geckos are by far the most vocal of all lizards and use both simple and complex calls to communicate with one another. Most other lizards are mute or produce only hisses or atonal squeaks or grunts. This is another legacy of geckos' nocturnality, as visual signals are not very effective at night. Having said this, however, geckos are also the only lizards—and indeed the only vertebrates—known to have good color vision at extremely low light levels. Geckos have a better sense of smell than other lizards, but their vomeronasal system (another chemical sense important for locating prey; see "How do geckos communicate?" in chapter 4) is not as well developed. As a consequence, they rely more heavily on vision for hunting. Because of this geckos, as a rule, rely more on ambush techniques for hunting than on active searching for prey based on chemical cues.

How many kinds of geckos are there?

As of November 2012 about 1,490 species of geckos have been recognized. This number changes on almost a weekly basis, as new geckos, as many as 54 per year, are described. The increase in gecko numbers in recent years reflects both active fieldwork by herpetologists in what had been poorly studied areas and advances in technology. The use of DNA sequencing to investigate relationships among different organisms can reveal the existence of previously unrecognized cryptic (hidden) species that may differ only subtly, or not at all, in morphology. It can also help systematists who describe and classify living beings confirm whether observed differences in the morphology (anatomy) are reflective of genetic divergence and can be reliably used as evidence for the recognition of new species. Geckos account for about one-fourth of all species of lizards. As their rate of description is among the highest for any lizard group, this proportion is likely to grow. Only skinks, with more than 1,500 species, outnumber geckos.

A few of the approximately 121 genera of geckos (the number depends on which of several competing taxonomic interpretations are followed) are monotypic, or having only a single species. These include the Festive Gecko (*Narudasia festiva*) of Namibia and the Long-necked Northern Leaf-tailed Gecko (*Orraya occultus*) of Australia. At the other end of the spectrum, a few genera of geckos are hugely species-rich. The four richest genera are the Dwarf Geckos (*Sphaerodactylus*) with 101 species of mostly leaf-litter dwelling forms in the Neotropics, the Forest Day Geckos (*Cnemaspis*) with 103 species, and two closely related genera, the Split-toed Geckos (*Hemidactylus*) and the Bent-toed Geckos (*Cyrtodactylus*), with 123 and 171 species, respectively. *Cyrtodactylus*, in particular, accounts for many of the recently described gecko species, having more than doubled in the past decade. This is a genus with a broad distribution from Nepal, Tibet, and north India to Australia and the Solomon Islands, and it seems to have speciated in association with disjunct rock formations, with almost every isolated limestone system in tropical Asia having an endemic species.

Gecko species are not uniformly distributed either phylogenetically—across the different families—or geographically. Families of geckos that have at least some species with adhesive toepads (Gekkonidae with 930 species,

The Festive Gecko (*Narudasia festiva*) of Namibia is the only member of its genus. Several such genera are similarly distinctive and have no particularly close relatives. Courtesy of Johan Marais.

The Bent-toed Gecko genus *Cyrtodactylus*, represented here by *C. consobrinus* from Sarawak, Borneo, is the most species-rich of all gecko genera, with approximately 170 species. Courtesy of L. Lee Grismer.

Diplodactylidae with 124, Sphaerodactylidae with 208, and Phyllodactylidae with 127) all have many more species than those that are entirely padless (Carphodactylidae with 29 species, Pygopodidae with 41, and Eublepharidae with 30), suggesting that the evolution of toepads may have opened up new ecological opportunities for the gecko lineages that posses them. With respect to distribution, Europe has the poorest representation, with only four species. The Americas with 250 species (40 percent of which are Dwarf Geckos, many in the West Indies) have only about as many kinds of gecko as Australia and the islands of the Pacific, despite their much greater area. Africa and Asia have the greatest number of species, but these are heavily clumped. In Africa the greatest diversity is in the Horn of Africa and the arid southwest of the continent, as well as in Madagascar, which alone has 100 unique species. In Asia more than 95 percent of species occur south of 40° N.

What is the current classification of geckos?

For a long time, the relationships between different groups of geckos remained poorly known. Geckos were largely classified by the structure of their feet. This allowed herpetologists to place geckos into many different genera, but did not help to determine how those genera were related to one another. The current classification of geckos is based on genetic data derived from DNA sequencing, but most of the major groups recognized in this way can also be distinguished by at least some shared morphological or anatomical characteristics.

The group that includes all of the living geckos as well as their closely related fossil relatives is referred to as the Gekkota. The Gekkota is, in turn, divided into a total of seven living families of geckos (the familial position of fossil geckos more than 50 million years old is uncertain). Three of these families, the Carphodactylidae, Diplodactylidae, and Pygopodidae, comprise the Pygopodoidea, whereas the other four, the Eublepharidae, Sphaerodactylidae, Phyllodactylidae, and Gekkonidae constitute the Gekkonoidea. Within the Pygopodoidea, the limbless pygopods are most closely related to carphodactylids. All of the pygopodoid families are limited in their distribution to the Australian region, with pygopods also reaching New Guinea (two species), and diplodactylids present in New Zealand and New Caledonia. Gekkonids are most closely related to phyllodactylids, and these families together to sphaerodactylids, whereas eublepharids are equally distantly related to all other gekkonoids. Gekkonids are mostly distributed in Africa and Asia, whereas the Phyllodactylidae and Sphaerodactylidae both have New World and Old World (North Africa through Central Asia) representatives. The eublepharids have what is probably a relict distribution—much smaller and more fragmented than it was in the past, with its few species scattered in North America, Africa, the Middle East, India, Japan, and southeast Asia.

What characterizes the major groups of geckos?

Three of the gekkotan families are easy to distinguish from all others. Pygopodids (also called pygopods) all lack forelimbs and have only small flaps that represent the hindlimbs. Despite being decidedly un-gecko-like in this regard, their eyes have a transparent covering called a "spectacle" like that of most geckos. They also share a mobile skull, which allows for a wider gape and improved force and control in grasping and biting, and other typical gekkotan traits. Carphodactylids and eublepharids both lack adhesive toepads that other limbed geckos have, and they both have leathery-shelled eggs. The root name *Eublepharis* means "true eyelid." Members

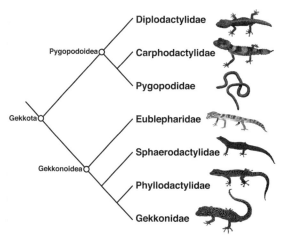

A phylogenetic, or evolutionary, tree of the seven living families of geckos. The virtually limbless Pygopodidae has evolved from within the families of more typically limbed geckos.

of this family are indeed the only geckos that retain eyelids and lack spectacles. Carphodactylids all have some sort of specialized tail structure: *Orraya*, *Phyllurus*, and *Saltuarius* have broad-leaf tails, *Nephrurus* has thick tails that end in a small knob, and the remaining genera have more-or-less carrot-shaped tails with thick bases and finely tapered tips. At least some members of the remaining families have toepads, and all have the transparent spectacle. Like the pygopodoids, members of the Eublepharidae have flexible, leathery-shelled eggs, whereas the remaining gekkonoids have the derived feature of rigid, calcareously shelled eggs.

These diverse families are difficult for a nonexpert to tell from one another and the characteristics that truly distinguish them are features of osteology and genetics. For this reason, the families Phyllodactylidae and Sphaerodactylidae were not recognized as separate from the Gekkonidae until 2008. Although these families diverged from one another in the Cretaceous and have been evolving independently since then, they have also been diversifying for a long time. As a result, there is as much morphological variation within these families as there is between them. In practice this is not really a problem, because genera of geckos are relatively easy to characterize on the basis of external features. Herpetologists know which genera belong to which families (see Appendix A). A combination of foot morphology and scalation details of the head, body, and tail are usually the features that are needed to correctly identify the correct genus of a gecko.

When did geckos evolve?

Lizards resembling geckos appeared during the Mesozoic Era, about 150 million years ago. It is rather difficult to say which of the few Mesozoic gecko-like fossils are true geckos, however. Geckos exhibit many features

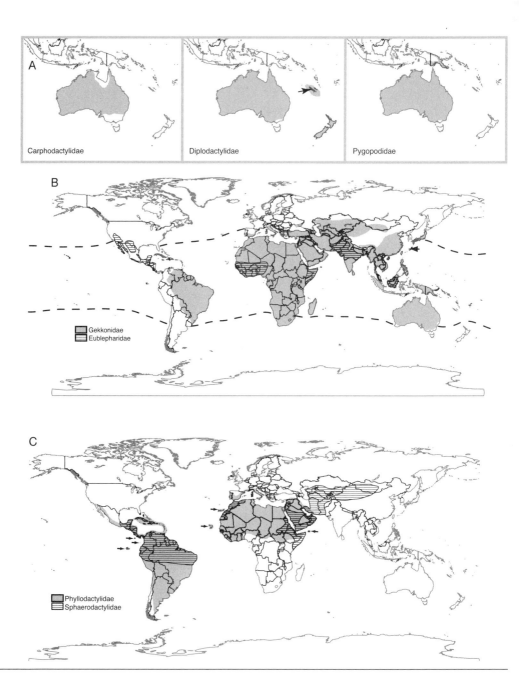

The global distribution of geckos by family. The three pygopodoid gecko families (A) are all restricted to Australia and surrounding landmasses, although none occur on Tasmania. In A, the arrow indicates New Caledonia. The eyelid geckos (Eublepharidae) have a relict distribution, that is, their 30 species scattered over the tropics and subtropics, mostly north of the Equator (B) are likely the remnants of a group that was once more widely distributed. The Gekkonidae are especially widely distributed in the Old World and have reached most island groups in warmer seas (bounded by the dashed lines in B). They have also become naturalized after being introduced in many areas of the Americas (introduced populations not shown). The arrow indicates the presence of eublepharids on the Ryukyu Islands of southern Japan. Both sphaerodactylids and phyllodactylids have transatlantic distributions (C). In the Americas, the former group is mostly tropical, whereas the latter extends almost to the tip of mainland South America. Arrows indicate islands occupied by these two families. Phyllodactylids occur on most of the islands, but only sphaerodactylids are on Cocos Island, Costa Rica (x), and Socotra Island, Yemen (y), is home to geckos of both families.

The Cat Gecko (*Aeluroscalabotes felinus*) exhibits the features typical of the Eublepharidae—movable eyelids and narrow, padless toes. However, its long slender limbs and thick coiled tail are specializations of this species alone. Courtesy of L. Lee Grismer.

The Banded Knob-tailed Gecko (*Nephrurus wheeleri cinctus*) from Western Australia shows several features typical of the Carphodactylidae, including a relatively large head and short, broad tail as well as narrow, padless toes. Courtesy of Brad Maryan.

that are primitive for lizards as a whole and the fragmentary remains of many fossils often do not preserve some of the most distinctive characteristics. Although not all paleontologists and systematists agree on which fossils are true geckos, several fossils from the Mid to Late Cretaceous have been described as geckos, but all of these are known only from incomplete specimens. However, *Eichstaettisaurus schroederi*, a fossil lizard from the Late Jurassic (150 million years ago) of Germany is represented by a complete skeleton, and it is very gecko-like. It has some features that are lost in all modern geckos, such as a parietal foramen, a small opening in the skull roof that allows light to reach the so-called third, or parietal, eye. It is otherwise similar in most respects to living geckos. The age of *Eichstaettisaurus* is consistent with estimates of the age of geckos based on other lines of evidence.

Molecular phylogenies, evolutionary trees derived from DNA data, may be dated if one or more genes that have been sequenced to produce the phylogeny evolves in a clock-like manner and if this "clock" can be calibrated using one or more reference points (known dates or ranges of dates, such as those of a fossil or of a geological event believed to have resulted in lineage splitting). Time trees derived from these data suggest that geckos split from other lizards at about 200 million years ago, at the beginning of the Jurassic Period and that the diversification of the living gecko groups began at about 140 million years ago, at the start of the Cretaceous. All seven living families were probably present by the end of the Cretaceous, 65 million years ago, but this cannot be confirmed by the meager fossil remains known at present.

What is the oldest fossil gecko?

Eichstetttisaurus schroederi is the only putative gekkotan from the Jurassic Period, but it is probably best considered as gecko-like, a member of the more inclusive group to which geckos belong, but not a true gecko itself. Several fossils from the Cretaceous have a better claim to being true geckos, even if they are not members of any of the living families. One of these is *Cretaceogekko burmae*, a fossil in amber that is 100 million years old. It has toepads, strongly suggesting that it is a gecko, but without skeletal information, it is impossible to say much more. Two conventional fossils represented by incomplete skulls from the Cretaceous have also generally been considered to be true gekkotans. The oldest is *Hoburogekko suchanovi* (about 110 million years old). This species has a tubular frontal bone like living geckos but has some other features that no other gecko shares. *Gobekko cretacicus* at approximately 83 million years old is much younger.

The Jurassic fossil *Eichstaettisaurus schroederi* comes from the same geological formation as the famous fossil bird, *Archaeopteryx*. Although it has many skeletal features that modern geckos lack (such as the parietal foramen, an opening in the skull roof) it is very gecko-like in appearance and is probably similar in body form to the earliest geckos.

Geckos: The Animal Answer Guide

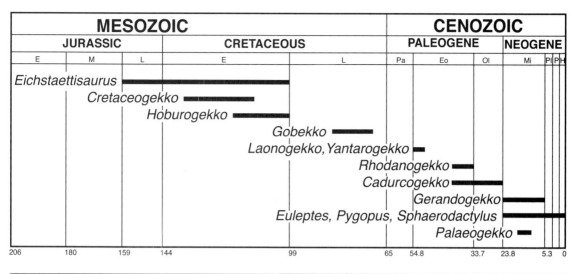

The distribution in time of some fossil geckos and gecko-like lizards. The scale below the diagram is in millions of years before the present. E, M and L designate Early, Middle, and Late. Abbreviations for periods within the Cenozoic are Pa = Paleocene, E = Eocene, Ol = Oligocene, Mi = Miocene, Pl = Pliocene, P = Pleistocene, and H = Holocene.

It is similar to living geckos in almost all aspects, but its frontal bones are paired and open below, an atypical condition for geckos. Both of these species are probably true geckos, but they may be representatives of families that did not survive to the present, which would explain their anatomical differences. It is likely that other fossils that would clarify the early history of geckos do exist in museum collections, but in fragmentary form. However, most microfossil accumulations await study and can be difficult to work with because of the small and delicate nature of the bony elements.

Where have fossil geckos been found?

Eichstaettisaurus specimens have been referred to two different species from the Jurassic and Cretaceous of Germany, Spain, and Italy. The Cretaceous gekkotans *Hoburogekko* and *Gobekko* both come from the Gobi Desert of Mongolia, whereas *Cretaceogekko burmae* lived in Burma (Myanmar) in tropical Asia. Gecko fossils from the Cenozoic Era are more widely distributed and include the Eocene *Yantarogekko balticus*, which was found in amber from the Baltic Sea coast in northwest Russia, and fragmentary material from the Paleocene of Brazil (unnamed), the Eocene of France (*Rhodanogekko, Cadurcogekko, Laonogekko*), and the Miocene of Europe and North Africa (*Gerandogecko, Palaeogekko*). The still-living *Euleptes* and *Sphaerodactylus* are also known from Miocene fossils from central Europe and the Dominican Republic, respectively, and a Miocene *Pygopus* has been found in Australia. Many Holocene fossil geckos are known, but these nearly all represent living genera or even living species and come from throughout the current range of geckos, wherever appropriate fossil-bearing deposits have been investigated. Some of the places where fossil geckos have been found, such as central Europe or the Baltic, experience long cold winters and are today devoid of geckos. However, the global climate has undergone many cycles of warming and cooling and at various times in the past conditions favored the expansion of tropical animals and plants toward the poles. Northern Europe in the Eocene, for example, was much warmer than at present and enjoyed a subtropical to tropical climate that supported not only geckos but also other lizards now occurring only at lower latitudes.

What did extinct geckos look like?

The gecko bodyform seems to have been a successful shape to be for a very long time. The Jurassic lizard *Eichstaettisaurus schroederi* suggests that the earliest geckos were similar in appearance to modern geckos, with large heads and large eyes and a large number of small teeth. Like most living geckos, they were probably somewhat dorsoventrally flattened. The earliest geckos did not have adhesive toepads and were similar to non-padded geckos and other lizards in having the fifth toe of the hind foot set off from the other toes. *Eichstaettisaurus* exhibits an autotomized (broken) tail, with a break present within the fifth caudal vertebra. This is the spot closest to the torso where tail breakage occurs in most living geckos and is anterior to where it is placed in most other lizards. Our best idea of the outer appearance of long extinct geckos comes from specimens that have been preserved whole in amber, a substance derived from the resin of trees. The sticky resin can trap plants or animals, which are eventually covered. Over time the amber hardens and the trapped organisms are then referred to as

Yantarogekko balticus is a 54-million-year-old fossil in amber from northwest Russia. Its preserved exterior reveals that geckos have changed little since the Eocene and its one preserved foot (*lower left*) shows that this tiny gecko was a climber like many of its living relatives.

Courtesy of Wolfgang Weitschat.

Geckos: The Animal Answer Guide

"inclusions." Only small animals are likely to be trapped in the resin, so vertebrate inclusions are rare compared with those of insects, spiders, or plant seeds or pollen. The oldest gecko in amber, *Cretaceogekko burmae*, is known only from part of tail and a leg, but this is enough to show that it had adhesive lamellae under the toes (see "How strong is a gecko's grip?" in chapter 2). *Yantarogekko balticus* is a 54-million-year-old gecko represented by most of a forebody in amber and demonstrates that Eocene geckos looked very much like their living relatives. Like most living geckos, it was covered with small granular scales, had large, lidless eyes, and relatively short limbs. Similar to two-thirds of living geckos, it also possessed adhesive toepads, but the configuration of these differed from that in any living species. Other geckos in amber are much younger, only 15 to 20 million years old, and belong to the living genus *Sphaerodactlyus*. Some such geckos are virtually intact and can even be assigned to extant species groups.

Chapter 2

Form and Function

What are the largest and smallest living geckos?

Geckos are mostly small-bodied lizards. The average gecko is only about 2.0–2.3 inches (50–60 millimeters) in head and body length, and less than 3 percent of geckos are larger than 5.5 inches (140 millimeters) in this dimension (Table 2.1). The largest living gecko is the New Caledonian Giant Gecko *Rhacodactylus leachianus*, which reaches a head and body length of 10 inches (256 millimeters). Several other geckos reach longer total lengths, but this is because they have much longer tails than *R. leachianus*. Although in the wild this species seldom weighs as much as 8 or 9 ounces (250 grams), captive individuals can reach 1.3 pounds (600 grams). Other very large geckos include the Tokay (*Gekko gecko*), Smith's Green-eyed Gecko (*Gekko smithii*), and the largest of the Madagascan Leaf-tailed Geckos (*Uroplatus fimbriatus* and *U. giganteus*), all of which can reach head and body lengths of more than 7 inches (180 millimeters). The Indian Leopard Gecko, *Eublepharis fuscus*, has been reported to reach nearly the size of the largest *Rhacodactylus*, but I have not seen a specimen approaching this size. The largest gecko that ever lived was *Hoplodactylus delcourti*. This giant from New Zealand is known from a single stuffed specimen that was probably collected in the early 1800s. Its total length is just over 2 feet (622 millimeters) of which 14.6 inches (370 millimeters) is head and body length. Not surprisingly for such a large lizard, it was known to the indigenous Maori people of New Zealand but was probably already rare by the time the first Europeans colonized the Pacific. To put the size of this species in perspective, its skull is longer than the average gecko's body. The largest pygopods (members of the genus *Lialis*) can reach even longer total lengths (31 inches, or 780 millimeters) than those of Delcourt's Gecko, with head and body lengths greater than *Rhacodactylus*. However, being limbless and slender, they are much less massive than these genera.

Several geckos vie for the title of smallest, but the record is probably held by the Jaragua Dwarf Gecko, *Sphaerodactylus ariasae*, that reach only 0.5–0.7 inches (14–18 millimeters) head and body length as an adult. Its relative, the Virgin Islands Dwarf Gecko (*S. parthenopion*) is about the same size. These miniature inhabitants of the leaf litter are not only the smallest of all lizards, but the smallest of almost 25,000 species of amniotes (see "What are geckos?" in chapter 1). The mass of the largest gecko is estimated to have been as much as 4.4 pounds (2 kilograms), or more than 15,000 times that of the smallest.

When are geckos active?

Geckos as a group almost certainly evolved as nocturnal creatures. This is reflected in their large eyes, ability to vocalize, lack of a parietal foramen, and in the relatively drab colors of most species. As one of only a few lizard groups to become nocturnal, geckos were able to capitalize on a lack of competition from other reptilian insectivores. This is believed to be one of the main reasons why geckos have become so diverse. Most nocturnal geckos become active around dusk or shortly after dark and may remain active for at least several hours. Depending on the air temperature, activity usually decreases by the middle of the night, but this is highly variable. In southeastern Australia Dtellas (*Gehyra variegata*) were found by herpetologist H. Robert Bustard to be most active for 3 hours after sunset. There are several reasons why activity is greatest earlier in the evening. Prey availability may be greater at this time, and light levels, although low, may still be favorable for visual hunting. The most important reason is that geckos have no way of maintaining or increasing their body temperatures

The New Caledonian Giant Gecko (*Rhacodactylus leachianus*) is the largest living species of gecko. It reaches over 10 inches (256 mm) in head and body length, but its prehensile tail is very short. Courtesy of Mark O'Shea.

Table 2.1. Maximum size (head and body length or snout-vent length [SVL]) of the largest species of geckos (140 millimeters/5.5 inches snout-vent length or greater). Most species occur on islands.

Species	SVL (mm)	SVL (inches)	Occurrence	Comments
Hoplodactylus delcourti	370	14.6	New Zealand	Extinct
Lialis jicari	310	12.2	New Guinea	Reduced-limbed
Lialis burtonis	290	11.4	Australia, New Guinea	Reduced-limbed
Rhacodactylus leachianus	255	10.0	New Caledonia	
Eublepharis fuscus	252	9.9	India	Length unconfirmed
Pygopus lepidopodus	240	9.4	Australia	Reduced-limbed
Pygopus nigriceps	227	8.9	Australia	Reduced-limbed
Pygopus robertsi	224	8.8	Australia	Reduced-limbed
Uroplatus giganteus	200	7.8	Madagascar	
Uroplatus fimbriatus	195	7.7	Madagascar	
Gekko smithii	191	7.5	Southeast Asia	
Rhacodactylus trachyrhynchus	190	7.5	New Caledonia	
Phelsuma gigas	190	7.5	Rodrigues (Mascarene Islands)	Extinct
Pygopus schraderi	180	7.1	Australia	Reduced-limbed
Gekko gecko	178	7.0	Southeast Asia	
Gekko reevesii	173	6.8	China, Vietnam	
Cyrtodactylus novaeguineae	172	6.8	New Guinea	
Eublepharis angrimainyu	170	6.7	Middle East	
Uroplatus lineatus	170	6.7	Madagascar	
Gekko albofasciolatus	165	6.5	Borneo	
Cyrtodactylus irianjayaensis	163	6.4	New Guinea	
Cyrtodactylus robustus	161	6.3	New Guinea	
Hoplodactylus duvaucelii	160	6.3	New Zealand	
Uroplatus henkeli	160	6.3	Madagascar	
Cyrtodactylus zugi	159	6.3	New Guinea	
Eublepharis macularius	158	6.2	Pakistan and adjacent areas	
Gehyra vorax	156	6.1	Fiji	
Gekko verreauxi	155	6.1	Andaman Islands	
Ailuronyx trachygaster	152	6.0	Seychelles	
Gekko siamensis	150	5.9	Thailand	
Cyrtodactylus tripartitus	148	5.8	New Guinea	
Mniarogekko chahoua	147	5.8	New Caledonia	
Blaesodactylus boivini	144	5.7	Madagascar	
Saltuarius salebrosus	143	5.6	Australia	
Cyrtodactylus klugei	143	5.6	New Guinea	
Cyrtodactylus salomonensis	142	5.6	Solomon Islands	
Gehyra georgpotthasti	142	5.6	Vanuatu/Loyalty Islands	
Gehyra marginata	142	5.6	Indonesia	
Saltuarius cornutus	141	5.6	Australia	
Gekko vittatus	140	5.5	Indonesia, Melanesia	
Rhacodactylus trachycephalus	140	5.5	New Caledonia	
Haemodracon riebeckii	140	5.5	Socotra	

The Jaragua Dwarf Gecko (*Sphaerodactylus ariasae*) from the Dominican Republic measures no more than 0.7 inches (18 mm) in head and body length and vies with another member of its genus for the title of smallest gecko. Courtesy of Tony Gamble.

at night. As ectotherms, they ultimately rely on an external source of heat, the sun. After nightfall, geckos and surfaces they move on all begin to cool down. Body functions like locomotion and digestion become less efficient at lower temperatures, so nocturnal geckos typical return to their retreat sites once temperatures drop below a certain level. For geckos in cooler climates, this may limit activity to just a few hours after dark. In warmer climates geckos can be active longer, but, if they have been successful in their search for food, there is little benefit in remaining exposed and so activity still declines well before dawn.

Many different groups of geckos, however, have evolved to become secondarily diurnal. This is especially true on islands where predators are absent or less numerous and, consequently, the cost to being more conspicuous may be lower. In the absence of high predation risk or strong competition from other diurnal lizards, geckos that become day-active are able to take advantage of the physiological benefits of higher temperatures. Geckos that have made this transition often have relatively smaller eyes and brighter colors than do their relatives as adaptations to higher light levels and a greater reliance on visual communication. The Green Geckos (*Naultinus*) of New Zealand and the Day Geckos (*Phelsuma*) of the Indian Ocean islands are good examples. Like other diurnal lizards, they need to wait until their body warms to a suitable temperature before they become active. Activity can continue throughout the day, but typically geckos spend much of their time motionless, although they may still be alert to their environment, particularly the movement of potential prey and of possible threats.

Form and Function

Although most geckos are nocturnal, some are crepuscular (active mostly around dawn and dusk) and others, like the Fan-toed Gecko (*Ptyodactylus guttatus*) can be active during both periods of night and day.

Courtesy of Tony Gamble.

Quite a few geckos are chiefly nocturnal but may also be active during part of the day. A well-studied example is the Fan-toed Gecko (*Ptyodactylus guttatus*). Other geckos are crepuscular (active around dawn and dusk). This may be a good strategy for some tropical geckos to avoid activity during the hottest periods of the day while still taking advantage of the availability of light to make foraging easier. The Dwarf Geckos of the Western Hemisphere (*Sphaerodactylus*) include diurnal, nocturnal, and crepuscular species.

Gecko activity is also affected by other environmental variables. For example, most, but not all, nocturnal geckos seem to increase activity when there is more moonlight available, presumably because this helps them see to forage for prey. Under certain circumstances, too much light may make geckos decrease activity to avoid exposure to their own predators. Wind and heavy rain also tend to decrease gecko activity, probably because they restrict the movement of arthropod prey or make them more difficult to locate or catch.

Geckos also show seasonal variations in their activity patterns. Activity is usually greatest during seasons of highest prey availability and during the mating season (for those species that do not breed year round). In areas that experience cold winters, a drop in temperature may initiate a period of inactivity called "brumation." In this state the gecko may not feed or move away from its retreat and its metabolic rate may be quite low, but it does not truly "shut down" as do animals that hibernate. On warm winter days brumating geckos may be able to take advantage of the favorable conditions and become active. However geckos in areas that experience very cold winters with months of freezing temperatures, such as *Alsophylax pipiens*, a species of the central Asian steppe, there may be no activity for as much as half of the year.

Geckos: The Animal Answer Guide

The Giant Day Gecko (*Phelsuma grandis*) is bright green with red markings. Such bright colors are found in geckos chiefly among diurnal arboreal species. Courtesy of Tony Gamble.

Do geckos sleep?

Yes, geckos do sleep—in a manner of speaking. Like other terrestrial vertebrates, geckos appear to require periods of rest. However, sleep in geckos and other lizards is different from that of mammals and birds, in which wakeful rest and true sleep are very different neurologically. Whereas birds and mammals have periods of REM (rapid eye movement) sleep during which dreams occur, this does not seem to be the case for reptiles. Nonetheless, they do experience restful periods during which they are not responsive to many sensory stimuli. These periods occur at night for diurnal geckos and during the day for nocturnal ones. In geckos with true eyelids (Eublepharidae), it is easy to tell when sleeping occurs because they close their eyes. However, in geckos with spectacles, the eyes cannot close and sleep can be harder to recognize. Sleeping geckos remain motionless, and the head is usually not held in an alert posture as it is when active. Geckos sleeping on the ground may curl their tails around themselves and pull the limbs in toward the body, but climbing geckos are capable of sleeping while holding onto vertical or even inverted surfaces. Many arboreal geckos often sleep in a head-down position with their tails out straight behind them.

Why do geckos have big eyes?

Geckos have large eyes, on average larger than any other group of lizards. This is related to their nocturnal history as a group, because larger eyes are advantageous for night vision. The size of the orbits in the skulls of some fossil gekkotans reveals that this was a feature of even the earliest members of the group. Eye size is greatest among terrestrial species of nocturnal geckos. Israeli herpetologist Yehudah Werner hypothesized that this may be to compensate for the ground-dwellers' more limited field of vision. Incidentally, he also proposed that the elevated posture that many ground geckos adopt is also related to increasing the field of vision. Diurnal geckos typically have smaller eyes than their night-active relative, but these are still larger than those of lizards in ancestrally diurnal relatives.

Can geckos see color?

Geckos have amazing eyes. Humans and other mammals have two types of receptors in their retinas: cones to see color in bright light and rods for mostly colorless vision in low light. Lizards have only cones and so generally have excellent color vision, partly explaining why so many lizards are brightly colored. Geckos also have only cones, but they have adapted these for nocturnal use, enlarging them so that they function more like rods and thus function well in dim light. Geckos have cones that are sensitive to blue, green, and ultraviolet light, but lack the cones that allow other lizards to see red. Gecko eyes also have large pupils that dilate from tiny pinholes in bright light to circular openings up to 300 times larger in the dark, allowing in as much light as possible. Together, these adaptations give geckos something that no other vertebrate is known to have: color night vision. In the Helmet Gecko, *Tarentola chazaliae*, light sensitivity is 350 times greater

Terrestrial geckos, like this Spider Gecko (*Agamura persica*), have the relatively biggest eyes of any gecko. Eyes like these have large pupils that let in as much light as possible for these nocturnal lizards. Courtesy of Tony Gamble.

Geckos: The Animal Answer Guide

than a human's at the lowest light levels at which each can see color. This means that geckos can discriminate different colors in even dim moonlight. Diurnal geckos like *Phelsuma* have much smaller cones and smaller eyes with circular pupils. This gives them color vision similar to that of other lizards, but means that they cannot see well at night.

Do geckos have eyelids?

Yes, all geckos have eyelids, but only in the 30 species of eublepharid geckos can they move. Eublepharids can close their eyes and blink just like humans. However, all other geckos have a structure known as a spectacle or brille that covers the eye. The spectacle is actually derived from a transparent lower eyelid that has become fused with the upper eyelid. The result is a transparent disk that is positioned directly in front of the eye itself.

The spectacle is continuous with the gecko's skin. When the gecko sheds its skin, the spectacle is shed as well. Some non-eublepharids, like the Barking Geckos (*Ptenopus*) and the Knob-tailed Geckos (*Nephrurus*), have eyelid-like structures called "extrabrillar fringes." These partly cover the upper portion of the eye and may be raised or lowered slightly, but animals with this structure cannot close their eyes. The spectacle is believed to protect the cornea beneath from abrasion and to reduce the rate of evaporative water loss from the large surface of the eye.

Why do geckos lick their eyes?

All geckos, including the limbless pygopods, use their broad tongues to lick or "tongue wipe" their faces and eyes. This behavior is stereotypical and occurs in the lidded eublepharids as well as in those geckos with

The eye of the Helmet Gecko (*Tarentola chazaliae*) has been shown to be capable of distinguishing color under light conditions similar to dim moonlight. Geckos' abilities to see in color at night are unique among vertebrates. Courtesy of Jean-François Trape.

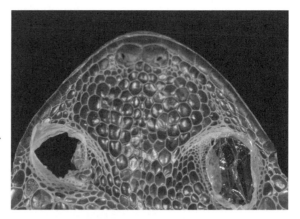

In most geckos, the eye is covered by a transparent "brille," or spectacle, that represents the fusion of the upper and lower eyelids. When geckos shed their skin, the transparent spectacle is also shed as can be seen on this FitzSimons' Gecko (*Chondrodactylus fitzsimonsi*). Courtesy of Bill Branch.

Some desert-dwelling geckos, like the Wonder Gecko (*Teratoscincus scincus*) have visor-like structures that shade the eyes, but these "extrabrillar fringes" are not true eyelids, which are present only in the eublepharid geckos. Courtesy of Bill Branch.

spectacles. It is thought that this has a cleaning function, which would be especially important in the absence of eyelids but is still useful for removing sand or other large debris even in geckos with eyelids. Geckos will also lick their face and eyes to "drink" if there are water droplets on their head. In addition, tongue wiping usually occurs after eating and sometimes after other strenuous activities like running. The function of this action under these circumstances is unknown.

What are the bulges on the necks of some geckos?

Many geckos develop large swellings along the sides of the neck. These are commonly mistaken for some sort of tumor or parasite, but they are, in fact, quite normal. These are structures known as "endolymphatic sacs." Found in other animals, including humans, these sacs contain fluids and are part of the inner ear. In most other lizards, this system lies entirely within the skull, but in gekkonoid geckos, as well as the diplodactylid *Eurydactylodes*, these sacs extend along the sides of the neck. In geckos, only species

The Web-footed Gecko (*Pachydactylus rangei*) of the Namib Desert uses its tongue to wipe the spectacle, presumably clearing it of debris as eyelids would otherwise do. However, even eublepharid geckos, who have eyelids, exhibit this behavior. Courtesy of Johan Marais.

producing rigid-shelled eggs (see "Why do geckos lay hard-shelled eggs?" in chapter 6) typically have such structures.

The endolymphatic fluid contains calcium that accumulates in the sacs as a crystalline form of calcium carbonate called aragonite. The sacs of some geckos become hugely inflated with these deposits. These are normally larger in females than in males, because they may possibly serve as reservoirs of calcium for rapid mobilization into the calcareous eggshell. In some juvenile geckos, they may also be enlarged, perhaps to serve as a source of calcium during the rapid periods of bone growth early in life. In captivity an excess of calcium can result in abnormally enlarged endolymphatic sacs. Although excess calcium in the diet can cause some health problems, the enlarged sacs themselves are not a danger to the animal. Species of Day Geckos, *Phelsuma*, seem to exhibit the most enlarged endolymphatic sacs of any gecko. Fat-necked Geckos from New Caledonia, *Oedodera*, have enlarged necks, but this is unrelated to the sacs.

Are some geckos limbless?

Most geckos have normallydeveloped limbs and five toes on each of their four feet. In some geckos, there has been a reduction in the length of digits, particularly the thumb (digit I), which is reduced to a nubbin in geckos like *Perochirus*, *Lygodactylus*, and *Phelsuma*. Pygopodid lizards (also called pygopods or flap-footed lizards), however, are effectively limbless geckos. Although older classifications treated pygopods as a family separate from limbed geckos, many lines of evidence now clearly demonstrate that they have evolved from limbed gecko ancestors. Pygopods are one of many lineages of lizards that have independently undergone a reduction in the limbs and an elongation of the body.

Calcium accumulates in the endo-lymphatic sacs of many geckos. These structures—also found in the heads of other animals, including humans—contains liquid used in hearing and balance. The accumulation of calcium produces bulges on the neck, particularly in females, where it may be mobilized for the production of the calcified egg shell, as in this Indian Bark Gecko (*Hemidactylus leschenaultii*). These bulges are often mistaken for tumors but are both harmless and normal in most geckos. Courtesy of Ashok Captain.

The forelimbs of pygopods are entirely lacking but the skeleton of the shoulder girdle is retained inside the ventral body wall and even a tiny rudiment of a humerus, the basal arm bone, is present is some species. All pygopods retain a pelvis and at least the basal bones of the hind leg, although these are very tiny. Digits are present in the skeleton of some pygopods, but externally the limb appears as a flap without any trace of separate toes. The flaps may be used to assist with locomotion on certain types of terrain. It has been proposed that males may use them in courtship, but this has not been confirmed. Many pygopods are mostly surface-active, but there are burrowing specialists, like *Ophidiocephalus* and *Aprasia*. Others live in and around grass tussocks where they move at some distance above the ground by "grass swimming," pushing the bends in their elongate bodies against the grass stems. Burrowers tend to have short tails and small eyes, whereas surface active species have long tails and larger eyes, although never as large as nocturnal limbed geckos. Pygopods are mostly Australian, with only two species reaching neighboring New Guinea. Thus, it is not surprising that their closest relatives are the carphodactylid geckos, a padless group of Australian lizards.

How do geckos climb?

If people know one thing about geckos, it is that some of them can climb up walls or hang onto a pane of glass. How they manage to do this was a topic of study for more than a hundred years. Contrary to popular belief, geckos do not have suction cups on their feet, nor do they produce some sort of sticky substance that allows them to hang on. The keys to climbing

Geckos: The Animal Answer Guide

With elongate bodies, no visible forelimbs, and only tiny flaps for hind limbs, pygopods, like this Eastern Scalyfoot (*Pygopus schraderi*), look nothing like other geckos. However, its spectacle, lightly built skull, and other features identify pygopods as highly modified geckos. Courtesy of Tony Gamble.

are tiny, hair-like structures called "setae." These are derived from even smaller, simpler projections, called "spinules," which are projections of the epidermis found on the scales of even non-climbing geckos. Setae range in length from about 20 to 100 microns (a micron—or micrometer—about 0.00004 inches) and may be complexly branched, with each branch ending in a small, flattened tipcalled a "spatula." When these minute tips are brought into contact with a surface that the gecko is walking on, two types of forces come into play. Shear forces result from friction between the setal tip and the surface. Van der Waals forces, based on tiny molecular level interactions, actually create temporary molecular bonds between the gecko and the surface. These forces are very weak at the scale of an individual setal tip. However, there are as many as a billion such tips on the foot of a gecko, all arranged in fields on the surface of scansors, enlarged scales on the under surface of the toes. The sum of all of these forces is enough to hold the gecko onto the surface it is climbing on.

Setae are necessary for gecko adhesion, but they are only part of a complex system that makes climbing possible. The gecko must have a way to attach and detach the setae and to control their placement. As many setal tips as possible need to come in close contact with the surface. This is no mean feat, as even smooth surfaces are really quite uneven and irregular at the scale of microns. Climbing geckos have either a fat pad or a system of small blood vessels in the toes that act to cushion the scansors and help them conform to the substrate to ensure that as many setal tips as possible actually reach the surface. Once the spatulate tips of the setae touch the surface, they must be placed in tension (either by the weight of the gecko or by muscles and tendons pulling on the scansors) for the adhesive forces to take full effect. To be detached, the setae must be peeled off the surface. This is done by hyperextending the toes, which causes them to roll off the surface from their tips backward. Of course, this whole process must be repeated rapidly and with great precision with each step. This involves not

Climbing geckos have expanded pads under their toes that allow them to adhere to most surfaces. These may be restricted to the tips of the toes or cover much of the digit, as in the East African Fragile-skinned Gecko (*Elasmodactylus tuberculatus*)

Courtesy of Bill Branch.

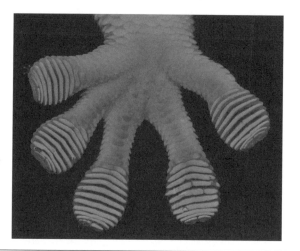

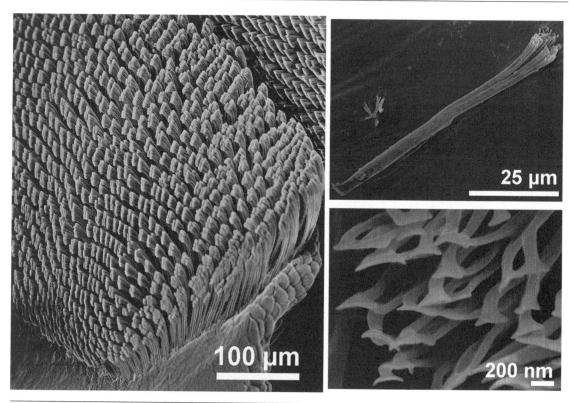

The scansors, or adhesive structures, of the Tokay Gecko (*Gekko gecko*) each support vast arrays of setae, tiny projections of the outermost layer of the epidermis (*left*). Each setal stalk (*above right*) is branched. and its many tips end in minute spatulae (*below right*), which interact with the surface when the animal walks, establishing adhesion through shear and van der Waals forces. Courtesy of Michael Bartlett.

only the skin and the musculoskeletal system but also touch-sensitive organs that give the gecko details about the position of each toe and scansor and parts of the ear that tell it about its body orientation. Some geckos, like the Dwarf Day Geckos, *Lygodactylus*, and most geckos in New Caledonia, also carry adhesive pads on the tips of their tails, allowing them to use the tails as a "fifth limb" when climbing. Most geckos also have claws on the toes, and these also assist in climbing, particularly on rough substrates like bark, where it may be hard for the scansors and their setal tips to conform to the surface. Indeed, claws make it possible for many geckos without adhesive scansors to be accomplished climbers nonetheless.

How fast can geckos run?

Most geckos are not as speedy as many other lizards, but some terrestrial geckos can run quite fast. The fastest gecko known is the Namib Day Gecko, *Rhoptropus afer*. It lives in parts of the desert that have flat rocky outcrops and, when disturbed or threatened, runs from rock to rock, often covering distances of more than 16 feet (5 meters) or about 100 body lengths in a single sprint. This species has been clocked at a speed 6.7 miles per hour (10.7 kilometers per hour). Most other geckos that have been measured can only run at 0.4–4.0 miles per hour (0.7–6.4 kilometers per hour). Geckos can only climb at about two-thirds of the speed that they can run because the attachment and detachment of the adhesive setae from the surface takes time and slows the gecko down with each step. However, they are just as fast on steeper inclines as on gentle ones. The added speed on horizontal surfaces is achieved by hyperextending their toes, thus holding the adhesive scansors off of the ground.

The fastest gecko is the Namib Day Gecko (*Rhoptropu safer*), which can reach speeds of 6.7 miles (10.8 km) per hour. Its long limbs and toes contribute to its sprint speed and its small scansors, carried at the tips of the toes, can be raised from the ground while running. Courtesy of Trip Lamb.

Can geckos run upside down?

The van der Waals forces that allow geckos to climb vertically also allow them to hang on and even run while inverted. However, to use friction, an important part of the adhesive mechanism, in this position, geckos must actively control their scansors to keep their setae in tension (in a vertical position gravity helps to do this). They must also exert more effort to keep their bodies close to the surface. In some geckos, particularly larger ones, this may not be possible. In buildings in the tropics it is not uncommon for house geckos to drop from the ceiling as they pursue insects. In part, how well geckos can run upside down or even hang vertically depends on the kind of surface they are on. Some surfaces, like the nonstick substance Teflon, which is used to line pots and pans, have a very low surface energy, meaning that it is difficult to establish van der Waals bonds. On these surfaces most large geckos cannot stick at all and even smaller geckos need to climb continuously to keep from sliding down.

What are flying geckos?

No geckos can truly fly, that is stay aloft using powered flight. However, several different groups of geckos have independently evolved the capacity to glide or parachute. Both gliding and parachuting involve morphological or behavioral adaptations that decrease the rate of descent of a falling animal. In gliding the angle of descent is shallow so that the animal can cover a long horizontal distance as it drops, whereas parachuters typically drop steeply, albeit at a slow rate. The most accomplished gliders, commonly called "flying geckos" are members of the genus *Ptychozoon* (meaning "animal with folds"). These are arboreal geckos found from India south and east to Borneo. They typically have pronounced flaps of skin along the sides of the body, extensive webbing between the toes, and a flattened tail with lateral skin flaps. Other geckos with similar but less strongly developed adaptations include some members of the genus *Luperosaurus* and two species of *Hemidactylus*. Skin flaps in other geckos, such as *Uroplatus* and *Rhacodactylus* are believed to function chiefly in disguising the animal's body outline rather than in locomotion, but members of both genera can and do jump regularly, and these skin flaps may provide some assistance. The Turnip-tailed Gecko, *Thecadactylus rapicauda*, has no body flaps, only webbing between the toes, but it is nonetheless capable of parachuting to move between trees in the Amazon. A large part of the ability to parachute seems to rely simply on adopting the correct posture when jumping, with limbs and toes spread, but both the tail and the toe-webbing play important roles in controlling the orientation of the animal in the air.

"Flying" geckos are actually para-chuters that slow their descent by using webbed feet and folds of skin along the edges of the body. Although members of several genera have this ability, it is most well developed in Southeast Asian geckos in the genus *Ptychozoon,* such as the Kuhl's Parachute Gecko (*P. kuhli*).

Courtesy of L. Lee Grismer.

How strong is a gecko's grip?

Geckos are theoretically able to hold on with a force that greatly exceeds that necessary to support their own body mass. Different studies have indicated that this amount is from several hundredfold to four-thousandfold or, for a 1.75 ounce (50 gram) Tokay Gecko, sufficient to support the weight of about three 150 pound (70 kilogram) adult humans! This overdesign is probably because the natural surfaces that they move on are really quite uneven and rough at the scale of the setal tips. This means that only a small proportion of the gecko's possible points of attachment will really be used as it climbs on a particular surface and that even this fraction of adhesive surface must be capable of supporting the entire animal. Further, at times in the gecko's stride only one foot, or even one toe, may be supporting the entire body. The amazing capacity of the gecko's grip has inspired humans to search for new and better artificial adhesives.

Can geckos hang on when they are dead?

Geckos can hang on to surfaces even after they die. Although geckos can control the movements of their toes very precisely, the basic component of their adhesive system depends on purely physical forces that can function even in a gecko that is unconscious or dead. As long as the gecko's

setae are clean and dry, shear and van der Waals forces will act to hold it to a substrate once the tips of the setae are in contact with a surface of even moderate surface energy. When the setae are in tension, these forces will act until the toetips are lifted and the toes hyperextended to peel the setae off the surface. In practice, the symmetrical feet of padded geckos distribute the animal's weight so that some setae will be in tension at all times. This capability may be proven by placing a dead gecko with its head pointing upward on a pane of glass or other material. In this position the setae will mostly be in tension and will easily support the body's weight. That this occurs in nature has been described by naturalists from the Mascarene island of Mauritius who reported that, following a cyclone, Day Geckos, *Phelsuma cepediana*, were found still attached to branches after having been killed in the storm.

Can geckos swim?

There are no truly aquatic geckos, but some species will occasionally enter the water to escape predators or to capture prey. One New Zealand species (*Dactylocnemis pacificus*) has been observed to dive into pools to escape danger, holding on to submerged stones to avoid floating to the surface. Most geckos will not only float if placed in water but will also swim quite handily simply by using their normal pattern of terrestrial limb and

Geckos: The Animal Answer Guide

body movements. Very small geckos, such as the tiny sphaerodactyls *Coleodactylus* and *Lepidoblepharis*, have water-repellent skin and are so light in weight that they do not break the water surface but rather walk across it. Even larger geckos, particularly those with prominent keeled scales, are capable of trapping a film of air around themselves and can also float on the surface. Geckos do not appear to purposely swim or float to travel from place to place, but it is probable that some insular geckos have crossed water barriers accidentally by means of currents. In the eighteenth century, when most biologists had never seen a live gecko, it was widely believed that species with extensive webbing between their toes, such as the Leaf-tailed Gecko of Madagascar, *Uroplatus fimbriatus*, were semi-aquatic and used their feet like paddles in the water.

Can all geckos lose and regrow their tails?

Tail autotomy (literally "self-cutting") is the ability of an animal to lose a portion of its tail, and this is usually associated with some regenerative capability. In lizards, autotomy is widespread and is dependent on the existence of an autotomy plane within the vertebrae. This plane is a preformed zone of weakness that permits the tail to break at this specific point when force is applied to it. The autotomy plane corresponds to a connective tissue sheet that separates adjacent muscle bundles along the length of the tail and sometimes to zones of weakness in the overlying skin and to valves in the blood vessels of the tail. Thus, when lizards autotomize their tails, there is little or no blood loss and all of the tissues break easily and without

Although no geckos regularly swim, small sphaerodactylids, like the Dwarf Gecko (*Coleodactylus brachystoma*), have water-repellant skin and are so light that they can walk on the surface film of the water when their forest floor homes are inundated.

Courtesy of Tony Gamble.

excessive trauma. The force that initiates the break usually comes from a potential predator or even a rival of the same species, but the lizard can "voluntarily" break the tail by initiating the violent contraction of the tail muscles while using even slight outside pressure on the tail as a fulcrum.

The main benefit of tail autotomy is as an antipredator strategy. If the attention of a potential predator can be distracted by the tail, autotomy allows the lizard the opportunity to escape, preserving its own life and its future ability to reproduce at the expense of a replaceable body part. Nonetheless, there are still consequences to the loss of the tail. In many fast-running lizards, the tail plays an important role in balance. Tail autotomy slows them down. In the Australian Marbled Gecko (*Christinus marmoratus*), however, tail loss has been demonstrated to increase running speed, presumably simply through weight loss.

Bibron's Gecko (*Chondrodactylus bibronii*) is like nearly all geckos in its ability to autotomize—literally self-cut, or break off—the tail as an escape mechanism. Muscular contractions cause the tail to break at preformed zones of weakness. The tail will eventually regenerate, although without vertebrae. Courtesy of Tony Gamble.

The Marbled Gecko (*Christinus marmoratus*) of temperate Australia loses energy stores when the tail is shed. It is able to run faster without its tail, unlike many non-geckos, in which the tail is proportionally longer and is used as a counterbalance. Courtesy of Tony Gamble.

Geckos: The Animal Answer Guide

Another cost faced by autotomizing geckos is energy loss. Many geckos have the capacity to store lipids in the tail as an energy reserve. In the Banded Gecko (*Coleonyx variegatus*), the loss of the full tail may result in a female's loss of a reproductive season as tail lipids are required to provide the energy needed to make egg yolks. Alternatively, individual eggs may be smaller and contain less energy, compromising the size and future success of hatchlings. In males, tail loss may result in a decrease in social status and potentially in reduced mating opportunities. In extreme cases the lack of lipid stores may increase fatality during periods of food scarcity. Geckos, therefore, typically autotomize as little of the tail as possible in order to escape would-be predators. However, in some cases, especially geckos with fat tails or broad leaf-shaped tails, such as *Uroplatus, Phyllurus,* and *Saltuarius,* either the anatomy of the tail or perhaps the functionality of it does not permit it to break at all points along its length. In these geckos autotomy is limited to the base of the tail, and tail loss is an all or none proposition. In only three species of geckos is autotomy not possible at all, even at the tail base. All are members of the genus *Nephrurus,* which typically have short broad tails with a small knob at the tip. In *N. asper, N. sheai,* and *N. amyae* the knob is still present, but the remainder of the tail has become greatly reduced, so that it is only a fraction of the body length. Functionally such a tail may be too small to distract a predator and anatomically the tiny vertebrae do not develop the planes necessary for autotomy to occur.

Autotomy occurs most easily at lower body temperatures. Presumably this is because geckos' ability to run and therefore escape is lowered when it is cold and tail loss becomes the only life-saving solution available. Not surprisingly, in such dire situations it is also more likely that the entire tail will be shed, thereby maximizing the possible distraction (and reward) for the predator and buying extra time for the gecko to move to safety.

Tail regrowth normally occurs in all geckos that autotomize. The tail stump is quickly covered by a layer of cells that protects the exposed surfaces. These then multiply forming a blastema, a mass of cells that will regenerate the tail. Under good conditions, the tissues that make-up the tail will begin to differentiate and grow within 2 or 3 weeks and within 9 to 18 weeks a fully functional tail is formed. Regenerated tails may grow nearly as long as the originals, but they are usually a little shorter. They also differ in having no vertebrae (instead, a flexible cartilaginous rod is grown) and less well-organized internal and external structures than the original. The original color pattern is often replaced by irregular markings and in some cases the regenerated tail is much thicker. One of the geckos most often kept as a pet, the Crested Gecko, *Correlophus ciliatus* (formerly *Rhacodactylus ciliatus*), often does not reproduce its tail if it is autotomized at the base. It is unclear if this is the result of non-autotomic damage (the tail is broken, but not at one of the preformed zones of weakness), but it is more common in captive geckos than in the wild.

Why do geckos shed their skin?

The skin of geckos and all other squamate, or scaly, reptiles (lizards and snakes) has a complex morphology that includes numerous layers of cells in the epidermis, the outer portion of the skin. New cells are generated at the bottom of the epidermis. As the cells age, they migrate upward toward the surface. As they do so, the cells fill with keratin, a tough, flexible protein. Keratin gives reptile scales their protective properties and also prevents water loss through the skin. However, by the time the skin cells are fully keratinized they are also dead. This outer, dead layer of skin needs to be replaced periodically in order to accommodate the growth of the gecko. Of course, before the old skin is shed a new skin has formed beneath so that the animal always has a functional epidermis. The act of shedding is referred to as "ecdysis." Because it is associated with growth, ecdysis occurs more frequently in younger animals as they develop or during times of the year when more growth typically occurs. Prior to ecdysis, the skin usually becomes dull or cloudy in appearance. Geckos may shed in large pieces or the entire skin may be shed at one time. In the latter case the gecko will often eat the shed skin, which has some nutritional value (eating the skin also hides a telltale clue to the gecko's whereabouts from predators). It is typical that the skin on the toes, which have a very complex surface structure, may be more difficult for the gecko to remove than that on the head or body. Normally the spectacle covering the eye is also shed along with the rest of the skin, but under some circumstances it is not. Failure of the spectacle to shed results in short-term vision problems for the gecko and ultimately can lead to blindness. In the wild, geckos can seek out the best conditions for undergoing ecdysis, but captive ones may experience dysecdysis (incomplete shedding) if such conditions are not provided. A common problem is dryness; an increase in humidity or soaking the skin in water will often loosen the attached shed.

At what temperatures are geckos most active?

Geckos live in a diversity of habitats and experience a wide range of temperatures. Compared with diurnal lizards of other groups, nocturnal geckos are mostly active at lower and more variable temperatures. However, gecko locomotor performance, and presumably other activities, are actually optimized at higher temperatures comparable to those of diurnal lizards. Thus, geckos' preferred body temperatures are quite different from

Growth is accommodated by periodic ecdysis, or molting, in geckos. Ecdysis may also dislodge some ectoparasites and renew the setae that allow geckos to adhere to surfaces. Bibron's Gecko (*Chondrodactylus bibronii*) sheds most of the skin at once and usually eats it to salvage nutrients. Courtesy of Johan Marais.

Most geckos live in the tropics, but some species have successfully invaded the cool temperate zones. The Black-eyed Gecko (*Mokopirirakau kahutarae*) of the South Island of New Zealand lives in alpine habitats and can be active at temperatures as low as 48°F (9°C). © Tony Whitaker.

the temperatures at which they regularly function. Not surprisingly, geckos from more tropical areas are generally less tolerant of lower temperatures and may require air temperatures of at least 79°F (26°C) to forage. However, tropical geckos often have a narrower range of preferred temperatures than do subtropical or temperate zone geckos. Most geckos decrease activity at lower temperatures and are less likely to move away from their retreat sites. Below about 57°F (14°C) most geckos, even in cooler climates, will not try to forage. The critical thermal maxima (the temperatures that geckos can stand under extreme stress) are often very high—over 104°F (40°C) in many species and up to 110.8°F (43.8°C) in *Diplodactylus conspicillatus*. Preferred body temperatures are typically 73–95°F (23–35°C) but can approach 104°F (40°C). Field active geckos, however, are mostly in the 68–86°F (20–30°C) range, with some even lower. *Mokopirirakau kahutarae*, the Black-eyed Gecko of New Zealand, is essentially an alpine species, occurring at elevations of 4,100–7,200 feet (1,250 to 2,200 meters) far south in the southern temperate zone. It has been observed active at temperatures as low as 48°F (9°C) and may even be active when snow is on the ground if air temperatures allow. *Pachydactylus rangei* of the Namib Desert is also active at cool temperatures. I have seen it and several other geckos of the Namib coastal fog belt active and apparently foraging at temperatures below 50°F (10°C). The Atlas Day Gecko, *Quedenfeldtia trachyblepharus*, has even been reported to be active at air temperatures below 37°F (3°C), but its body temperature was probably significantly higher due to basking.

Chapter 3

Gecko Colors

What colors are geckos?

Geckos come in almost all possible colors. The most common colors are shades of brown, often light and dark in some combination of spots, blotches, stripes, reticulations, or bands. Gray, olive green, and black are also found. These drab colors serve geckos well, as they blend in well with the tree trunks, stones, and soil on which they are active. Many desert geckos have pastel earth tones to match the local sands. Especially beautiful reddish-browns or pinks are seen in some Knob-tailed Geckos, *Nephrurus*, and patterns including pale yellows are seen in *Coleonyx variegatus*, the Banded Gecko of the American southwest.

Bright colors are mostly limited to diurnal geckos. Green coloration appears in the patterns of several species, such as the New Zealand Harlequin Gecko (*Tukutuku rakiurae*), but it is truly characteristic of two genera, the Green Geckos of New Zealand (*Naultinus*) and the Day Geckos of the Indian Ocean (*Phelsuma*). Bright green geckos are always diurnal, and they are usually active on vegetation. Geckos are capable of seeing blue and green, so these colors allow them to use visual signals to communicate with one another. At the same time, green coloration serves as excellent camouflage against a background of vegetation, protecting geckos from birds and other visuallyoriented predators.

Among the most striking colors seen on any gecko must be the electric blue of the Tanzanian Dwarf Day Gecko, *Lygodactylus williamsi*. Bright yellow or red heads are characteristic of males of species that use visual displays, like many *Gonatodes*. Colorful patches on the throat are typical of some highly vocal species like the Bell or Barking Gecko, *Ptenopus garrulus*.

Gonatodes daudini, a small gecko from the Grenadines in the Lesser Antilles, has a striking combination of green and red in its color pattern and perhaps most spectacular of all is the aptly named *Cnemaspis psychadelica*, which has a yellow head, lavender body, and orange limbs, flanks, and tail. How geckos themselves perceive these colors is not entirely known. Although their eyes have retinal cones to detect blue and green wavelengths of light, the receptors most sensitive to reds are lacking. However, markings that we perceive as red may still be seen by geckos, but perhaps in the ultraviolet, for which geckos do have receptors.

In addition to the rainbow of naturally occurring colors in geckos, there is a diversity of "abnormal" conditions of coloration that can occur and may even be favored in captive animals. An example of such a condition is albinism, or the absence of pigment. This condition yields white or pinkish animals, the latter due to the visibility of blood vessels through the pigmentless skin. In geckos amelanistic albinism, in which melanin is not produced but other pigments may be present, is most common. Other conditions include melanism—the overexpression of melanin (giving brown or black skin), hypomelanism—a reduction in the normal amount of melanin, and xanthism—yellowish skin due to high concentrations of xanthophores (see "What causes the different skin colors of geckos?" below). These exceptional colorations also occur in nature, but many are likely to be deleterious. Geckos exhibiting them probably rarely survive to maturity, as they may be more conspicuous to predators. In some geckos, especially the frequently kept Leopard Gecko, *Eublepharis macularius*, breeders have developed "designer morphs" that exhibit colors or color patterns not normally occurring in the wild.

What causes the different skin colors of geckos?

The skin of geckos and other lizards contains several types of pigment-containing cells called "chromatophores." Deepest in the skin are "melanophores," which contain a brown pigment that yields a dark brown or black color. This is overlain by a layer of "iridophores" (also called "guanophores"). These do not contain true pigments, but rather are filled with plates made of guanine that are reflective and produce iridescence when they are illuminated. Most superficial are yellow-pigment-containing xanthophores and cells containing red or orange carotenoids called erythrophores. The interplay between the different layers of chromatophores ultimately gives the gecko its final color. In drab-colored geckos, the melanophores are responsible for the browns or blacks one sees. In yellow species, the xanthophores dominate. Green coloration is an interesting case, because there are no green pigment cells in geckos. Instead, the combination of bluish reflected light from the iridophores and yellow from the

Captive Leopard Geckos (*Eublepharis macularius*) come in a variety of artificially selected color morphs, including amelanistic albinism as seen here. All of the natural (and unnatural) color s seen in geckos produced from different combinations of just a few types of pigment-containing cells called "chromatophores" in the skin. Courtesy of Tony Gamble.

The colors of diurnal geckos like the Northland Green Gecko (*Naultinus grayi*) of New Zealand provide excellent camouflage. Green color is produced by the interaction of yellow xanthophores with reflected bluish light from iridophores. © Tony Whitaker.

xanthophores gives the animals a green color. Yellow individuals of the normally green *Naultinus elegans* sometimes occur in nature. These are the result of the absence or paucity of iridophores on their guanine plates. The conditions of albinism and melanism, among others, are the result of the abnormal under- or overexpression of the skin pigments.

It has been determined from captive-bred Leopard Geckos that some aspects of color can be modified by incubation temperature of gecko eggs. In these geckos, temperature during early development determines the sex of the animals, but incubation temperatures after this point can influence color. Animals raised at higher temperatures tend to be brighter and have less melanin, whereas darker animals with more melanin can be caused by incubating the eggs at a lower temperature. Even after hatching, keeping Leopard Geckos at cooler temperatures can result in darkening.

What color are a gecko's eyes?

Eye color varies in different species of gecko and sometimes even within a single species. Diurnal geckos viewed in daylight have large round pupils. Their eyes often appear dark from a distance, although closer inspection reveals the color of the surrounding iris. An exception is the Grenadines Gecko, *Gonatodes daudini*, in which a bright red iris adds to the gaudy colors of the body. In contrast, the pupils of nocturnal geckos close down to a narrow slit or series of pinholes in strong light, and the color of the iris is quite apparent. Geckos often have a grayish, beige, or brownish iris, although silvery or coppery colored eyes are also common. Some species, like Smith's Green-eyed Gecko, *Gekko smithii*, Nutaphand's Red-eyed Gecko, *G. nutaphandi*, or the Red-eyed Bent-toed Gecko, *Cyrtodactylus erythrops*, have distinctive bright eye colors that stand out against the drab body coloration. In many geckos, the appearance of the iris is strongly influenced by the complex patterns of the blood vessels that serve it. These are most often reddish or blackish and can form a dense net covering the surface of the iris. Black eyes, red eyes, and even eyes that are half colored and half black have been developed in Leopard Geckos through selective breeding.

Is there a reason for specific patterns on a gecko's skin?

The color patterns of most nocturnal geckos are probably best interpreted as a means of concealment. Many geckos have incredible camouflage, with color patterns that resemble the bark or stone surfaces on which they live. Particularly striking are leaf-like patterns in *Uroplatus phantasticus* and its relatives and the pale patches on some *Gehyra vorax* and *Mniarogekko*

The large eyes of geckos are striking. The iris is often whitish, silvery, or bronze, although many other colors are possible. In bright light the pupil of the Short-fingered Gecko (*Stenodactylus doriae*) is reduced to a narrow slit with a series of pinholes and the pattern of the blood vessels in the eye is clearly visible. Courtesy of Bill Branch.

Geckos: The Animal Answer Guide

chahoua that are uncannily lichen-like. Even geckos without such superb background-matching may be using color pattern to disguise themselves. Bands, stripes, and blotches can all be mechanisms to break up the shape of the gecko so that it goes unnoticed by predators that hunt using visual cues. Many geckos share certain elements in their color patterns like a banded tail, a pale collar around the neck, and a dark stripe through the eye. All of these serve to conceal the real outline of the body or to mask parts of the body that would otherwise be evident to a predator.

Especially for day-active geckos, many of the particular patterns seen serve as a means of communicating their species identity. For example, many places in the Indian Ocean have more than one Day Gecko species. Most *Phelsuma* have reddish and bluish markings on a green background. The particular arrangement of these different colors identifies members of a given species at a distance. Indeed, the large and relatively recent radiation of this genus probably was driven by the rapid evolution of color-based recognition systems. Other patterns may serve a role in courtship or in sexual recognition. In these cases the male and female geckos have different color patterns.

Are male and female geckos colored differently?

Even color night vision as good as a gecko's does not work particularly well at long distances. So, nocturnal geckos rely more on their other senses to identify potential mates. As a consequence, most night-active geckos do not show significant differences in male and female color patterns. However,

Many geckos, like this Madagascan Leaf-tailed Gecko *(Uroplatus* aff. *phantasticus)*, use coloration to conceal themselves, so patterns that blend in well with leaf litter, rocks, or other natural substrates are common. Courtesy of Miguel Vences.

there are exceptions. Males of species of the southern African genus *Chondrodactylus* always have bright white spots on their backs, adorning a pattern that is otherwise exactly like the female's. The role of these spots is uncertain, but it is probable that they are more easily visible (perhaps in the ultraviolet) at night than is the rest of the mostly brownish body. Because sexual differences in color pattern are uncommon in nocturnal species, even some prominent herpetologists have mistaken males and females of *Chondrodactylus angulifer* as two different species.

In many diurnal geckos, males and females are also the same color. This is true in the gaudy *Phelsuma*, as well as in the mostly gray *Rhoptropus*. In other day-active genera, like, the Neotropical genus *Gonatodes* and the Paleotropical *Cnemaspis*, some species exhibit a sexual color difference (dichromatism), whereas others do not. For example, *G. humeralis* males have bright red-and-yellow markings on the head, but the female is relatively drab. In these geckos, as in many other animals, because the female has a large investment in reproduction she cannot afford to make herself conspicuous and thereby risk drawing the attention of predators. Males vie

Some geckos, mostly diurnal species, exhibit some degree of sexual dichromatism—usually with a more brightly colored male and a duller female. In the Striped Neotropical Day Gecko (*Gonatodes vittatus*), the boldly striped male uses his color to send signals but increases his risk of being seen by a predator. Courtesy of Tony Gamble.

Geckos: The Animal Answer Guide

with one another for control of territories or access to females and may need to be more visible, in spite of the risk.

In *Ptenopus*, some *Cnemaspis*, and certain other geckos, males have brightly colored throats, even if body coloration is the same in both sexes. Such bright markings are common in especially vocal species in which visual and auditory cues may be important for reinforcing each other, yielding a stronger signal to others of the same species.

Do a gecko's colors change as it grows?

Many geckos keep the same basic pattern they had as juveniles, but this often becomes more faded as the animal grows. Brightly contrasting light-and-dark rings on the tail may become more muted and less conspicuous in the adult. This can be seen in the Leopard Gecko, *Eublepharis macularius*, in which the white bands in a young juvenile were so bright that they were mistaken for being bioluminescent, resulting in the description of a new species of glowing gecko! In some other geckos, like *Christinus marmoratus*, juveniles have a reddish tail, which fades in adults. In other species pattern elements are actually accentuated with age. For example, brighter colors may not develop fully until sexual maturity and in many geckos new pattern elements emerge with time. Many hatchling Leopard Geckos are relatively simply banded, but spots and blotches on the bands develop as the animals approach maturity. In the case of sexually dimorphic species, the juvenile males typically resemble the female in color, but change before they reach breeding age. Some of the most striking changes take place in some of the African geckos of the genus *Pachydactylus*. In *P. oreophilus*, the hatchlings have a boldly contrasting pattern, with a solid blackish body, pale hips, hind legs, and tail, and a white collar around the neck. As the geckos grow, their patterns change entirely and the adults have no trace of the original blackish torso. Such color changes can be important in helping scientists identify particular species. In the *Pachydactylus serval* group, adults of several species are nearly identical in appearance, but their hatchlings are distinguishable from one another based on color.

Can geckos change color?

Although they are not as accomplished as chameleons and none can change to match their background exactly, some geckos do change color. Most geckos can at least lighten and darken in response to temperature, light, and "emotional" state. Warm geckos, in the dark, that are not stressed are usually pale. Cold animals, in light, or in a disturbed or stressed state tend to be darker. This change can happen over a period of minutes (or seconds

A striking case of color change with age occurs in the Kaokoveld Gecko (*Pachydactylus oreophilus*) of Namibia. The boldly banded black-and-white hatchlings (*above*) grow into less strongly patterned adults (*below*), although hints of the juvenile pattern are still visible. Courtesy of Johan Marais.

in some cases) to hours. Some geckos appear whitish, pale pinkish, or almost transparent in their palest state. In many others, the full complexity of their patterns can only be seen in this paler state, whereas the darker phase obscures pattern elements and can make even a brightly colored gecko appear drab. An extreme occurs in *Pachydactylus bicolor*, a gecko from Namibia. In its dark phase, it is a chocolate brown with small yellowish or brownish flecks, but when it lightens a complex pattern of bands and blotches emerges. Males of at least some sexually dichromatic *Gonatodes*, like *G. humeralis*, rapidly change color from drab to brightly colored in the presence of rival males, however, the females remain drab all the time and are incapable of color change. These changes can be triggered by hormones or by signals from the nervous system. Change in state of health can also change color; for example, a heavy parasite load can dull the colors of normally bright males.

Is there much geographic variation in a single gecko species?

This topic has been little studied and is confounded by the fact that many widespread geckos, when studied in detail, turn out to be complexes of several different species. In other words, what appears to be one species turns out to be several similar species. So, what might have been interpreted as geographic variation within a species turns out to be variation among species. Geographic variation has been studied in the Speckled Gecko *Pachydactylus punctatus*, in southern Africa, but it now appears that this may also represent such a species complex. However, there is ample evidence for significant color variation, whether geographic or not, in many geckos. This is best seen in some of the most popular captive species. Leopard Geckos, *Eublepharis macularius*, for example come in a staggering diversity of color

Geckos: The Animal Answer Guide

Yellow-headed Dwarf Day Gecko (Gekkonidae: *Lygodactylus luteopicturatus zanzibaritis*); East Africa.
Courtesy of Bill Branch.

Golden Gecko (Gekkonidae: *Calo-dactylodes aureus*); India. Courtesy of Ashok Captain.

Sind Comb-toed Gecko (Gekkonidae: *Crossobamon orientalis*)**; Pakistan and India.**

Fat-tailed Gecko (Eublepharidae: *Hemitheconyx caudicinctus*)**; West Africa.** Courtesy of Tony Gamble.

Green-eyed Gecko (Gekkonidae: *Gekko smithii*)**; Southeast Asia.**
Courtesy of L. Lee Grismer.

Wonder Gecko or Frog-eyed Gecko (Sphaerodactylidae: *Teratoscincus scincus*); Southwest Asia. Courtesy of Bill Branch.

Psychedelic Rock Gecko (Gekkonidae: *Cnemaspis psychadelica*); Vietnam. Courtesy of L. Lee Grismer.

Viper Gecko (Gekkonidae: *Hemidactylus imbricatus*); Pakistan. Courtesy of Tony Gamble.

Web-footed Gecko (Gekkonidae: *Pachydactylus rangei*); Namib Desert, Southwestern Africa. Courtesy of Johan Marais.

Crested Gecko (Diplodactylidae: *Rhacodactylus ciliatus*); New Caledonia. © Tony Whitaker.

South Island Green Gecko (Diplodactylidae: *Naultinus tuberculatus*); New Zealand. © Tony Whitaker.

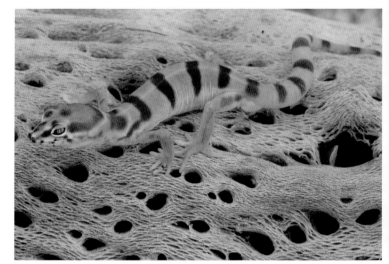

Western Banded Gecko (Eublepharidae: *Coleonyx variegatus*); **Southwestern United States and northwestern Mexico.** Courtesy of L. Lee Grismer.

Harlequin Gecko (Diplodactylidae: *Tukutu kurakiurae*); **New Zealand.**
©Tony Whitaker.

Thai Leaf-toed Gecko (Gekkonidae: *Dixonius siamensis*); **Southeast Asia.**
Courtesy of Tony Gamble.

Bogert's Flat Gecko (Gekkonidae: *Afroedura bogerti*); Northern Namibia and southern Angola.
Courtesy of Johan Marais.

Eyespot Gecko (Sphaerodactylidae: *Gonatodes ocellatus*); Tobago, West Indies. Courtesy of Tony Gamble.

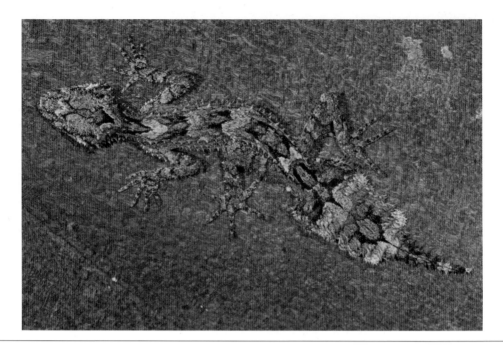

Southern Leaf-tailed Gecko (Carphodactylidae: *Saltuarius swaini*); Eastern Australia.

Courtesy of Jeff Wright and the Queensland Museum.

Banded Bent-toed Gecko (Gek-konidae: *Cyrtodactylus pulchellus*); Southeast Asia. Courtesy of L. Lee Grismer.

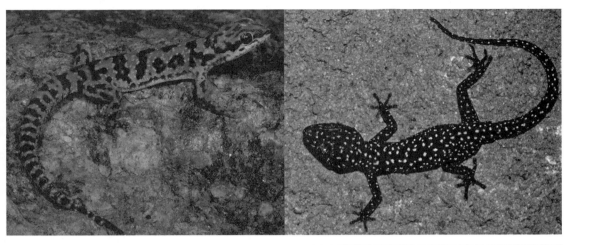

The Bicolored Gecko (*Pachydactylus bicolor*) of Namibia changes from a complexly banded and blotched light pattern to a dark pattern of bright white flecks on a dark brown background. (*Left*) Courtesy of Johan Marais. (*Right*) Courtesy of Mirko Barts.

Gargoyle Geckos (*Rhacodactylus auriculatus*) from New Caledonia come in a variety of striped, banded, and unicolored forms, all occurring at the same locality. In many geckos there is at least some variation in color and pattern across the distributional range. Courtesy of Josh Snyder.

patterns, including many "designer morphs" that have been bred over many captive generations and do not actually occur in nature. Even if some patterns are artificially created, this range of variation implies that the genetic potential for color pattern diversity exists. In the Crested Gecko, *Correlophus ciliatus*, and especially the Gargoyle Gecko, *Rhacodactylus auriculatus*, quite a range of colors and patterns exist in the wild. Although there may be some underlying geographic trends, many of the pattern variants in the latter species can be found at a single site among genetically nearly identical animals. Supposedly *R. leachianus* from individual tiny offshore islands in New Caledonia can be reliably identified by color pattern. This probably results from fixed differences in the populations inherited from tiny founder populations.

Chapter 4

Gecko Behavior

Are geckos social?

Some gecko species are quite social and others live mostly solitary lives. At a minimum, geckos need to be social enough for males and females to find one another and to mate. All geckos use some combination of vocalizations, visual signaling, and chemical signaling to communicate with one another (see "How do geckos communicate?" below). Males use these mechanisms to advertise their presence to females. When a receptive female is located, a courtship routine that is quite stereotyped is initiated (see "How do geckos mate?" below). Even the all-female species *Lepidodactylus lugubris* engages in pseudocopulatory behavior, with one gecko playing the male role and the other the female, although, of course, no genetic material is exchanged.

The same signals that are used by males to attract females are also used to keep other males away. For gecko species that occur in low density, same-sex social interactions may be limited to the occasional vocalization or fight when one gecko wanders too close to another. However, in close quarters social interactions become more numerous and more important. *Hemidactylus* geckos have been especially well studied from the viewpoint of social interactions, chiefly because some species live around humans and are, therefore, easy to observe. *H. turcicus*, for example, maintains dominance hierarchies within which both males and females have a certain status. This type of social structure seems to apply in cases where geckos are territorial. Dominant males will chase away subordinate ones but are usually tolerant of the presence of multiple females. Females may or may not tolerate one another. The unisexual geckos *H. garnotii* and *Lepidodactylus ligubris* have no dominance hierarchy but will fight other geckos when they are approached, defending only themselves, rather than a particular resource.

Under certain circumstances geckos will aggregate for specific benefits. For example, both heat and humidity can be maintained at higher levels by geckos if they are in close contact inside an enclosed space. *Underwoodisaurus milii* and *Rhacodactylus trachycephalus* both appear to use this strategy, piling up in heaps inside sheltering spots. Crevice-dwelling geckos like *Phyllurus platurus* in Australia will sometimes occur in large numbers, apparently taking advantage of particularly favorable sites. I have observed two cases of extraordinary aggregations of geckos, both involving well over 100 individuals. The first was the more-or-less terrestrial *Woodworthia maculata* on Stephens Island in New Zealand, where a mass of individuals were piled on top of one another under a sheet of tin. In the second example approximately 200 *Chondrodactylus* of two species, *C. turneri* and *C. bibronii*, and both sexes were under a dinner table–sized vertical rock slab in northern South Africa. I suspect in both cases that the thermal benefits of the sites drew the animals together. This is especially interesting in the latter case as males of these geckos are normally quite aggressive toward one another. So, they must have a mechanism to "turn off" this aggression when it is beneficial to do so.

The Golden Gecko of India, *Calodactylodes aureus*, may be one of the most social geckos. These geckos live in and around boulders in south India and use both sound and visual signals to interact with one another. In any group of geckos usually only one adult male, presumably the dominant individual, will assume a bright yellow color. If that male is removed from the population, another male will take on its color and apparently its social role. Pair bonding of a male and female has been suggested in the South American gecko *Gymnodactylus*, but the evidence for this is circumstantial. Viviparous geckos seem to show at least some maternal behavior when their babies are born (see "Do geckos care for their young?" in chapter

Unisexual geckos, like the Fox Gecko (*Hemidactylus garnotii*) will defend themselves in social interactions but do not defend territories or resources as do many bisexual species.

6) and "family groups" may be formed by adults and their offspring, but whether these geckos cooperate in some way or merely tolerate one another is uncertain.

Do geckos fight?

Yes, many geckos will fight with members of their own species or with other species under certain circumstances. Males of most geckos studied are territorial and will defend a particular resource, like food, a basking spot, or a shelter. They will not tolerate other adult males in their own territories and will engage in aggressive behaviors and even fight if intruders do not leave. However, males are usually tolerant of females. Females of some species will be aggressive toward other females, but others are not. Interestingly, at a single site in Australia *Gehyra variegata* was found to have no aggression between females, whereas *G. dubia* females would not tolerate others of their own sex. *Lepidodactylus lugubris*, an all-female species will also fight with one another. In many species, juveniles may be tolerated by adults but are in risk of being eaten if resources are low. Even geckos that are not really territorial in the strict sense may fight to retain access to resources. This may be seen around artificial lighting where geckos congregate to take advantage of the insects attracted to the lights. Under such circumstances, most geckos will resort to fighting in order to keep a favored position near the food as it flies or crawls in. When different species of geckos try to use the same resources, interspecies fights may occur, with larger and more aggressive species having the advantage. More often than not, however, larger species dominate smaller ones by means of competitive exclusion, essentially using intimidation rather than actual combat to keep smaller species from a food source. For captive geckos, it is critical that the social structure of the species being kept is considered and that situations that can lead to fights be avoided.

Do geckos bite?

Wild caught geckos occasionally bear the teeth marks of other geckos; these are most easily visible on the delicate ventral skin of the belly. Biting may be a regular part of territorial combat. In the giant *Rhacodactylus leachianus*, tree hollows for nesting and for sheltering may be a scarce commodity. Both male-male and female-female combat to obtain and maintain control over such sites can occur. Biting is common, and with these large geckos serious wounds can be inflicted. Tails may be bitten off. The loser in battles may be quite bloodied. Fights to the death probably rarely, if ever, occur, as greatly mismatched smaller geckos generally avoid combat with obviously larger and stronger rivals. I have observed several *R. leachianus* in the field with an entire foot or even limb missing and the stump nicely healed. This may have been the result of a failed predation attempt, but it might also have been the outcome of a lost territorial battle.

Although entirely subjective, I can provide some perspective on gecko bites, having been bitten by several hundred species. Regarding bites to humans, most geckos are too small to inflict any damage whatsoever. A large percentage is too small to even get their jaws around anything except perhaps the webbing between fingers (and even this is too much for smaller sphaerodactylid geckos). Most average-sized geckos deliver a minor nip that neither breaks the skin nor leaves a mark. However, geckos bigger than about 3.15 inches (80 millimeters) head and body length can sometimes give a mildly painful bite, especially if they take full advantage of their mobile skulls and really bite down hard. In such cases, it is not the teeth that are the problem, but the sheer force of the bite, rather like being pinched by a pair of pliers. The champions in this middleweight division are *Chondrodactylus turneri* and the very similar *C. bibronii*. In my experience, these species are among the most prone to bite. When they bite they always bite as hard as possible, flexing the upper jaw and withdrawing the eyes in their sockets (see "Why do geckos eyes sink in when they bite?" in chapter 7). However, I have only had blood drawn on a few occasions and then only a drop or two if the gecko bit me on a particularly tender spot.

Next on the list would be *Rhacodactylus auriculatus*. This species tends to bite and let go more quickly than some others, but it has long, blade-like teeth and in this case it is the teeth, not the bite force that does the most damage. Although not particularly inclined to bite, this gecko can easily break the skin and cause a good deal of pain. *R. leachianus* is even less likely to bite, but when it does the huge jaws bite down with terrific force and the

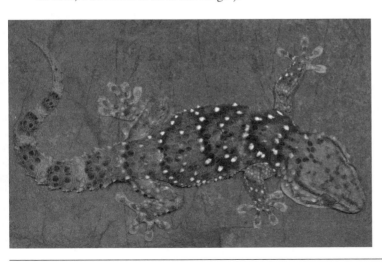

Turner's Gecko (*Chondrodactylus turneri*) is a relatively large gecko that can deliver a painful and tenacious bite. Individuals often bear the scars of previous encounters with others of their species, yet at times large numbers of individuals may be found together, perhaps taking advantage of limited sheltering or egg-laying sites. Courtesy of Johan Marais.

teeth, though small and peg-like are still big enough to do some damage. I have twice experienced this species draw blood and do not recommend the experience to anyone. Finally, the king of biting must certainly be the Tokay Gecko, *Gekko gecko*. Tokays are both lightning fast in their strike and impossibly tenacious in their grip. I was once bitten by a Tokay and worked hard for more than 20 minutes to remove it. If Tokays were the size of cats, no one would be safe. Like most geckos, Tokays usually bite down with more force if the prey (or your finger) pulls to get away. Tapping them on the snout usually does nothing, and attempts to lever open their mouths risk injury to the gecko and are usually unsuccessful. After 30 years of working with geckos, my best advice is to let the animal put all four feet on the ground and put it near something it will interpret as shelter, like a dark box, and then wait for it to let go.

How smart are geckos?

By mammalian standards, geckos are not very smart. As squamate reptiles, they lack the capacity for many of the higher brain functions that mammals have. But intelligence in the human or mammalian sense is probably not terribly relevant to geckos. Many of the behaviors of geckos are instinctive and allow them to carry out most of their life functions quite well. However, many experiments have shown that geckos can and do learn. Both positive and negative rewards have been successfully used to teach geckos to perform simple tasks, such as navigating through a maze or choosing between two different colors. There is also some evidence that geckos can remember some learned behaviors over time. The Tail-squirting Gecko, *Strophurus williamsi*, was found to have some means of returning to the same winter retreat site more than 80 feet (25 meters) from its summer shelter. This suggests that spatial information learned in one year was remembered in subsequent years. People who keep geckos as pets often teach their animals without even trying. Many geckos will learn that certain actions are always associated with food and will behave accordingly; for example by going to a particular spot in their terrarium in "anticipation" when their owner approaches.

Do geckos play?

Probably not, but it is hard to know for certain. Play behavior is well documented in birds and mammals, but in reptiles it has been tentatively identified in only a few species, mostly turtles, under captive conditions. Even in these cases, it was difficult to distinguish play from other types of exploratory behavior. Scientists generally believe that factors such as high

Geckos: The Animal Answer Guide

metabolic rate, endothermy (the ability to generate heat internally), and extensive parental care, which provide animals with surplus resources are usually necessary for play to evolve. Geckos explore their environment and may be "curious" about unfamiliar objects or organisms. However, most of this curiosity is almost certainly related to basic life functions. In other words, geckos want to know things like "Can I eat it?" "Will it try to eat me?" or "Can I mate with it?" Or, in the case of inanimate objects "Can it help or hinder me from eating, mating, or hiding?" Although there is a tendency for some people to interpret the behavior of their pet geckos as play, other more practical explanations probably apply.

How do geckos communicate?

Most night-active geckos communicate with each other by vocalizing. Geckos are the most "talkative" of all reptiles, and most species are capable of making at least some sounds. Gecko vocalizations can be divided into distress calls and multiple click (or multiple chirp or cluck) calls. Distress calls can be shrill and squeaky and designed to frighten or startle predators, or they may be clicks that communicate to other geckos that there is a threat. These calls can be made by both males and females. The Giant New Caledonian Gecko, *Rhacodactylus leachianus*, makes a loud growling sound in response to being grabbed by a human, predator, or rival, but it can also make a whistle, which seems to be more of an indication of milder distress. Multiple click calls are made most often by males and are a primary means of gecko communication because they are in the frequency range that geckos can hear. Such calls are used to attract mates or to help maintain territories or dominance hierarchies. Females of some species produce response calls to these.

The sound of these calls varies among species, with larger geckos generally producing calls of lower pitch. It also varies with temperature, with warmer geckos having a higher calling rate. The *to-kay* call of *Gekko gecko* is a familiar sound in parts of tropical Asia and gives this gecko its common name. Southern African Barking Geckos (*Ptenopus*) call mostly in the late afternoon and evening. In spring hundreds of males can form huge choruses, much like those of frogs. These can be so loud that the nineteenth-century zoologist and explorer Sir Andrew Smith stated:

> In the localities in which it occurs many individuals may be seen peeping from their hiding-places any time during the day, each uttering a sharp sound, somewhat like *chick chick*; and the number thus occupied is at times so great, and the noise so disagreeable as to cause the traveller to change his quarters.

> *Illustrations of the Zoology of South Africa*, Appendix, p. 6 (1849)

Geckos are the most vocal of all lizards. The Barking Gecko (*Ptenopus garrulus*) may be the most vocal of all geckos. Males, which have yellow throats, call to females from their burrow entrances, forming large choruses, like those of frogs. Courtesy of Bill Branch.

Diurnal geckos rely more on visual signals than sound to communicate. Carter's Semaphore Gecko (*Pristurus carteri*) has a laterally compressed tail, which is further exaggerated by a fringing crest of spines. The tail can be curled upward and moved about to send signals to other members of the species. Courtesy of Tony Gamble.

Day active geckos are less vocal than their nocturnal relatives. They often rely on visual communication. Many diurnal geckos are brightly colored and use these colors to draw the attention of members of their own species. Males of some species have brightly colored heads and bob these up and down to declare their presence. The Semaphore Geckos, *Pristurus*, wave their tails, which can be laterally compressed to increase their surface area and thus visibility, in different patterns to signal members of their own species. Other behaviors that are used in gecko communication include mouth snapping and push-ups.

Geckos also use chemical communication, although the details of how it works are poorly understood. Although both true smell and vomerolfaction (another method for detecting chemical stimuli in reptiles and other animals) are probably involved, the latter is probably most important in communication. Males (and rarely females) of many geckos have a series of pores in front of their vents or cloacas or on their thighs. These pores produce a waxy plug that contains chemical cues identifying the species and sex of the gecko and perhaps other information related to age, size, or state of health. The secretions are rubbed on the surfaces where the gecko walks and provide a marker of its presence to other geckos in the neighborhood. During courtship, geckos often lick one another. This is probably a form of both tactile and chemical communication. Molecules picked up by the tongue and transferred to the vomeronasal organ provide the courting geckos with information about their potential mates.

How do geckos make noise?

Geckos are among the only reptiles to have well-developed vocal cords, which allow them to make a variety of sounds. The vocal cords are folds of elastic tissue that lie inside of the cartilaginous larynx, or voice box. The larynx is attached to the lungs via the trachea and opens up into the mouth cavity through the glottis. When making one of its calls, the gecko passes expelled air from the lungs past the stretched vocal cords at a right angle. This causes the cords to vibrate and produce sound. At the same time the glottis is held open allowing the sound to enter the mouth cavity. The size, shape, and rigidity of the glottal opening, along with movements of the vocal cords, the position of the larynx in the throat, and the rate and timing of the expulsion of air from the lungs all play a role in particular type of sound a gecko will make. Some geckos also have modifications to the trachea that seem to enhance aspects of the call. Geckos can also make simple hissing noises, usually as part of defensive displays. These use the same structures as a true call, but a continuous stream of air is passed through the larynx without the stretching and vibration of the vocal cords or modification of the airflow by the glottis. One group of geckos, the Wonder Geckos or Frog-eyed Geckos, *Teratoscincus*, can also make noise using an entirely different mechanism. These have large overlapping scales along the top of their tails that bear tubercles both above and below. When rubbed against

Tokay Geckos (*Gekko gecko*) produce their characteristic *to-kay* calls by passing air through the larynx and over the elastic folds of the vocal cords. Air then exits through the glottis into the mouth. This open-mouthed gesture may be used in calling, but it also serves as a threat display. Courtesy of Tony Gamble.

one another as the tail is waved from side to side, these scales produce a rattling noise that is used as a warning. The Knob-tailed Geckos, *Nephrurus*, can also produce a sort of buzzing noise by vibrating their tail tips in dry leaf litter.

Do geckos have good hearing?

Yes! As one would expect, given that they have strong voices, geckos also have good hearing so that they can receive auditory information both from members of their own species and from other relevant sources in their environment. Geckos, including pygopods, have well-developed external ears. The ear opening is usually large, although it may be partly occluded by scales or skin folds. Like most other tetrapods, airborne vibrations strike a membrane, the tympanum or eardrum, and this, in turn, vibrates, transferring the movement to a cartilaginous extracolumella and thence to the slender stapes or columella, which in geckos is the only bone of the middle ear. This then vibrates against the oval window, which is the opening in the bony capsule that holds the inner ear. Here the vibrations cause movement in the fluid (endolymph) inside the inner ear. The movement of this fluid stimulates hair cells that trigger a nerve impulse that carries to the brain information about the nature of the hair cell displacement that can be interpreted as sound. These hair cells lie in a structure called the "basilar papilla," which in geckos is especially complex. Some exceptions occur in burrowing pygopods, like *Aprasia repens*, where the external ear and even the stapes can be lacking. Presumably these reductions reduce the sensitivity of the ear at higher frequencies. This is a common phenomenon in burrowing squamates, for whom low-frequency sounds and ground-borne vibrations are probably particularly important.

Geckos that have been studied have shown the greatest sensitivity to sound in the range of about 400 Hz–7 kHz (400 Hz is roughly the frequency of a Middle G on a piano, and 7 kHz is the frequency of the highest-

With the exception of some burrowing pygopods, all geckos have well-developed ears, like this Striped Leaf-tailed Gecko (*Uroplatus lineatus*) from Madagascar. The ear opening lies behind the angle of the jaws. The tympanum, or eardrum, is only slightly recessed from the skin. Courtesy of Tony Gamble.

Geckos: The Animal Answer Guide

pitched songbirds). However, it has been discovered that pygopods in the genus *Delma* are also sensitive to much higher frequencies, in the range of 12–14 kHz at a sound pressure of 75 decibels (the high end of the range for a loud speaking voice). This is the highest hearing limit known for any reptile. In contrast, most adult humans cannot hear sounds above 16 kHz (like the faint high-pitched whine produced by some electronic devices).

Just like their voices, gecko ears are also responsive to temperature changes and are more sensitive at higher temperatures, leveling off near the animal's activity temperatures. Thus a sound of a given frequency, in order to evoke the same auditory response at a cooler temperature would need to be louder (have more energy). Most geckos have a good match between their own voices and their ability to hear, so it is clear that gecko vocalizations can really be used for communication. However, some sounds made by geckos may be out of the range of their own hearing. Growling sounds, for example, may be aimed at potential predators with good hearing in lower frequency ranges.

How do geckos avoid or escape predators?

Geckos have an arsenal of techniques for avoiding predators or escaping if captured. Most geckos lead inconspicuous lives, spending much of their time in rock cracks, burrows, leaf litter, or tree holes and so avoid many predators by staying out of sight. Nocturnal geckos, which are most often drab in color, are further protected from many would-be predators by the cover of darkness. Geckos that bask or rest in the open rely on remaining hidden by "crypsis," using camouflage or other methods to avoid being seen. Many geckos in the genera *Rhacodactylus*, *Uroplatus*, and *Saltuarius*, for example, have skin flaps and folds that lie flat against the surface of trees, disguising the lizard's body outline. These and many others are well camouflaged, even having patches that look like lichens or moss. It has been suggested that a few geckos gain protection by mimicking other animals. The pygopod, *Pygopus nigriceps*, has a color pattern similar to that of juvenile *Pseudonaja nuchalis*, a highly venomous elapid snake with which it co-occurs. When the pygopod is threatened, it lifts up its forebody in snake-like fashion, widens its neck, and strikes putting on a rather convincing show. Several geckos, including those in the genera *Teratoscincus*, *Coleonyx*, *Eublepharis*, and *Cnemaspis*, curl their tails over their backs, perhaps mimicking scorpions.

If discovered, many geckos immediately freeze in place. This is an effective strategy when dealing with visually oriented predators that cue in on movement. For other predators, however, different strategies are necessary. If there is enough distance between the gecko and its attacker, fleeing

The Australian Thick-tailed Gecko (*Underwoodisaurus milii*) exhibits a characteristic defensive posture with its back arched and its legs straightened to make it seem as large as possible to potential predators. This sort of bluff is often successful. Courtesy of Brad Maryan.

The ability to lose skin to a predator to escape (much like tail autotomy) has evolved in several groups of geckos. The Southern Fragile-skinned Gecko (*Pachydactylus kladaroderma*) can recover from the loss of a great deal of its skin, although the physiology of this escape strategy remains poorly known. Courtesy of Bill Branch.

is always an option, particularly if a burrow, crevice, or other shelter is handy. Some pygopods use a special bounce or jump called a slide push that not only moves them away from danger quickly but also undoubtedly confuses the predator. The Australian Thick-tailed Gecko, *Underwoodisaurus milii*, like many terrestrial geckos, arches its back and straightens its legs to stand as tall as possible and hisses and lunges, trying to convince attackers that it is a larger, more dangerous opponent than it really is. At the same time, it slowly waves its tail from side to side, probably to distract the predator.

Geckos: The Animal Answer Guide

Other geckos squeak or bark a distress call or expose a brightly colored mouth lining that can startle predators. Some larger geckos, like the Tokay in Asia and Turner's Gecko (*Chondrodactylus turneri*) in Africa, can bite savagely and, if they cannot run to safety, will not hesitate to counterattack a predator. Australian geckos of the genus *Strophurus* and the New Caledonian *Eurydactylodes* have glands within the tail that secrete a sticky, somewhat unpleasant smelling substance that can distract predators or possibly physically prevent them from carrying out an attack. In the case of the former genus, this material can be squirted and aimed to strike at a distance. Geckos can also employ autotomy of their tails as a mechanism for distracting predators and gaining a chance for escape (see "Can all geckos lose and regrow their tails?" in chapter 2).

Perhaps the most extreme antipredator mechanism used by geckos is regional integumentary loss (RIL). This is essentially autotomy of the skin and should not be confused with normal skin-shedding (ecdysis), which all geckos do periodically. RIL is practiced only by some geckos, chiefly those that are the prey of certain types of predators that must handle or manipulate their prey before delivering a killing bite. Geckos with RIL have zones of weakness, like the perforations between stamps, that allow the skin to tear easily when pulled. These zones extend through nearly the entire skin so that all of the epidermis and most of the much thicker dermis can be pulled away. This leaves the gecko with only a thin layer of tissue left, which serves as a site for the regrowth of the skin. As with tail autotomy, the predator is startled and confused by RIL and is left with just a piece of skin to eat while the gecko escapes. The skin regrows over time, with larger wounds taking longer to heal. Up to 40 percent of the skin on the back can be lost, and the gecko can still recover. There are probably high costs to this behavior, however, as without the intact epidermis the gecko is susceptible to high rates of water loss and possibly open to infection.

Chapter 5

Gecko Ecology

Where do geckos sleep?

When sleeping or resting, geckos try to find places that will shelter them from potential predators and protect them from temperature extremes or bad weather. Terrestrial geckos usually shelter in burrows underground or beneath stones or fallen logs. Some, like the Australian Fat-tailed Gecko, *Diplodactylus conspicillatus*, use their tails to plug burrow entrances. Rock-living geckos retreat to cracks or crevices. Depending on the species, these may be shallow or deep, horizontal or vertical, or limited to particular types of rock. Many geckos prefer vertical cracks that open downward because they are harder for some predators to climb up and because they are protected from wind and rain and will not accumulate dirt or debris. Arboreal geckos often sleep beneath bark or in the axils of leaves, although they may also sleep in the open on branches, trunks, or leaves, particularly if they have some sort of camouflage. The live-bearing New Caledonian geckos, *Rhacodactylus trachycephalus*, sleep in family groups in tree holes. Pygopods that live in spiny grasses will generally sleep or rest deep in the grass tussocks. Even some arboreal geckos will come to the ground to seek out shelter in dense vegetation at the base of bushes. The choice of a sheltering site can also depend on other factors. Since geckos store heat they obtain from their surroundings, sheltering under thin rock flakes or bark may be especially advantageous, as these will warm quickly in the sun and transfer heat to animals sleeping beneath them.

Arboreal geckos often sleep head downward, stretched out along a tree branch or trunk. Here a well-camouflaged Gargoyle Gecko (*Rhacodactylus auriculatus*) rests on a lichen-covered trunk (*center*) in New Caledonia. This animal has a small radio transmitter on its back used to study its movements. Courtesy of Josh Snyder.

Which geographic regions have the most species of geckos?

Given that most geckos prefer warm climates, it is not surprising that geckos are most species-rich in tropical and subtropical regions. The country with the most gecko species is Australia, with more than 175, but there are also many geckos in Madagascar (more than 100), India (>90), southern Africa (South Africa has more than 80 and Namibia has more than 60), southeast Asia (Vietnam, Thailand, and Indonesia each have 60 or more species), and Iran (65). Some of these areas have moist forested habitats, but many are mostly drier and semiarid to arid. On the whole, more geckos live in deserts than in rainforests. Partly because of this, places like tropical South America have fewer geckos than might be expected, with the most in Venezuela (36) and only 35 in all of the huge territory of Brazil. Grasslands and savannas have among the fewest gecko species, probably because these more homogeneous, open habitats do not provide much three-dimensional diversity in microhabitat types. They also are poor in sheltered sites needed by geckos to thermoregulate, lay eggs, and rest. Likewise, desert dunes are also quite gecko-poor for just the same reasons.

Do geckos burrow?

Certain terrestrial geckos do occupy burrows. In some cases these may be abandoned by other small animals, such as spiders, scorpions, or other

This ground-dwelling Australian gecko, *Diplodactylus klugei*, is one of more than 175 species of geckos that live in Australia. The combination of a variety of arid habitats and the radiation of four different families of geckos in Australia contribute to its great diversity. Courtesy of Brad Maryan.

lizards, but generally they excavate their own burrows or at least modify existing ones. Geckos that burrow usually have modified feet to help them dig. Minimally their toes have reduced adhesive scansors, as these structures are not useful on sand or soil and would become clogged by dirt and sand grains. In extreme cases, the toepads are lost entirely. The African Giant Ground Gecko, *Chondrodactylus angulifer*, has lost the toepads of its ancestors and instead the skin of the feet bears small raised spines.

Other burrowing geckos, like the Wonder Geckos or Frog-eyed Geckos, *Teratoscincus*, of the deserts of Asia, have fringes on the toes that are used like rakes to help move sand as the animals dig. Even more specialized are the webbedfeet of the Namib Desert–dwelling *Pachydactylus rangei*. In this species, there are no adhesive pads and the feet are completely webbed to form small shovels for excavating of sand. Not only is there skin linking each toe to the next but the toes also have extra skeletal elements called "paraphalanges" that help to support the webs and provide attachment points for muscles controlling its movements. The Web-footed Gecko first pushes sand back with its front legs and then scoops it up and moves it away from the burrow entrance with its hind legs. They live in sandy dry river beds and in the dunes of the Namib. Although they may wander all over the dunes in search of food, their burrows are usually dug into the windward side of the dunes, because only there is the sand compact enough to keep the burrows from collapsing.

The Web-footed Gecko may have the most specialized structures for digging, the champion burrow-builders are the southern African Barking Geckos, *Ptenopus*. They excavate burrows that have a main tunnel as much as 2 feet (60 centimeters) long and 15 inches (38 centimeters) deep and may have several side chambers. These chambers often extend to just below the surface.In an emergency (like the invasion of the burrow by a predator) the gecko can quickly dig its way out. The male Barking Gecko usually

Geckos: The Animal Answer Guide

calls from the entrance of the burrow, which serves as both as an amplifying chamber to intensify the sound of the call and a retreat from predators alerted by these vocalizations. Despite the often-inhospitable daytime conditions on the surface, the burrow environment provides favorable temperature and humidity conditions.

How do geckos survive in the desert?

Deserts pose challenges of both temperature extremes and lack of water. Most desert geckos are nocturnal and escape high daytime temperatures by staying in burrows, rock crevices, or other retreats during the day. By indirect basking under rock flakes or near their burrow mouths, they are able to absorb sufficient heat to fuel them through their foraging period after dark, when temperatures can fall precipitously. Of course, not all deserts are hot all the time. In the coastal Namib Desert, where cool temperatures and heavy fog are common, geckos like *Rhoptropus bradfieldi* have become diurnal and evolved dark colors to help them absorb as much heat as possible during sunny periods. The night-active *Pachydactylus rangei*, *Colopus kochii*, and *Ptenopus carpi* can live in this fog belt as well; they tolerate low temperatures and are able to forage for food and search for mates when most lizards would be nearly immobilized by cold.

Desert geckos rarely if ever have access to free water, so most of them must obtain all the water they need from the body fluids of the prey that they eat. Here the geckos of the Namib fog belt have an advantage, as they have access to water that condenses from the fog onto all kinds of surfaces, including their own bodies.

Deserts are also quite limiting in terms of food resources. For animals like birds and mammals, dense populations can usually survive only in oases or other areas with adequate water. Geckos, in contrast, can thrive under these harsh conditions. As long as shelter is available, they can make

Two different strategies for geckos to burrow are to use fringed toes to rake or shimmy through the sand or to use webs as sand shovels to excavate. These approaches are used by Koch's Barking Gecko (*Ptenopus kochi*) and the Web-footed Gecko (*Pachydactylus rangei*), respectively. (*Left*) Courtesy of Bill Branch. (*Right*) Courtesy of Johan Marais.

Carp's Barking Gecko (*Ptenopus carpi*) lives in the harsh environment of the Namib Desert. It avoids the heat of the day by sheltering in burrows but must endure cold nighttime temperatures. It obtains water from the condensation of fog onto its own body and other surfaces.

Courtesy of Randall D. Babb.

do with the meager food supply the desert offers. A large part of this has to do with ectothermy (obtaining heat from external sources rather than generating it physiologically themselves). Because geckos do not need to use metabolic reserves to maintain a high body temperature, they use much less energy than do endotherms. If they do not need to move around, and especially if they remain at cooler temperatures, geckos can live on a very low energy budget. Even during times of plenty, many geckos do not feed every day and can go long periods without any food at all. When they do feed, their prey items can often be found taking advantage of the same shelters the geckos use, Other desert prey, like termites, are clumped in the environment and require geckos to expend more energy searching for nests. However, once located, a termite nest can provide a gecko with virtually limitless supply of food.

How do geckos survive the winter?

Geckos are not capable of withstanding freezing temperatures. When water in their cells freezes, it expands, destroying the cells and ultimately killing the gecko. Most temperatures above freezing can be tolerated by geckos, but below some threshold temperature, which varies among species, gecko function is impaired. Nearly all metabolic processes function most effectively at a particular range of temperatures. Because geckos are often active at temperatures below those optimal for many functions, they may not always be able to run at maximum speed or forage with the greatest efficiency, but they can nonetheless successfully carry out their life functions. Below their normal activity temperatures, however, the ability of geckos to move or to respond to stimuli drops off steeply. At such temperatures, an exposed gecko has no ability to flee or protect itself from predators. Thus, geckos become inactive and remain in their shelters when their body temperatures drop too low. Through basking or other behavioral means,

Geckos: The Animal Answer Guide

geckos can maintain their bodies above the ambient air temperature, at least for short periods (see "Do geckos bask?" below). For some truly tropical species 68°F (20°C) may be too cold. For cool-adapted geckos, activity may continue until the body temperature drops as low as 46–50°F (8–10°C; see "At what temperatures are geckos most active?" in chapter 2).

In reality, many geckos worldwide do not live in areas that drop to the freezing point or even to the point where body functions are seriously impaired. Nonetheless, some live in climates that do have "real" winters, either at high elevations or high latitudes. These geckos must find winter shelter where the temperature will not drop below freezing. Because the gecko's metabolism drops to a very low level at cold temperatures, feeding is not necessary during the winter, although overwintering geckos need to be well hydrated before becoming inactive. In the most extreme climates in the Northern Hemisphere, geckos will begin their quasi-hibernation known as "brumation" as early as October and may not emerge again until April.

Do geckos bask?

Many reptiles bask in the sun to raise their body temperature to some preferred level so that they can carry out activities like foraging and reproduction. Basking reptiles generally orient their bodies so as to maximize the solar radiation they are receiving. Otherwise, they remain quite still, although they may still be very observant of their surroundings. Like many other lizards, diurnal geckos also bask in order to raise their temperatures. Basking is especially critical for diurnal geckos in cold environments. The high-elevation Moroccan gecko *Quedenfeldtia trachyblepharus* can be active at elevations of up to 13,000 feet (4,000 meters), where snow may be present for much of the year. It takes advantage of every opportunity to bask, appearing from its hiding places whenever the sun comes out and disappearing as rapidly when clouds close in. To maximize the heat it can absorb, this gecko is blackish in color. The Namib Day Geckos, *Rhoptropus bradfieldi*, use a similar strategy. Although their environment is not as cold, temperatures along the foggy coast where they live are still well below the optimum for gecko physiology. In addition to basking and using dark skin to absorb as much heat as possible, they also gain heat from the black boulders on which they bask.

Most nocturnal geckos do not bask directly. Rather they absorb heat through contact with warm surfaces, a strategy called "thigmothermy." Geckos often choose their resting sites by maximizing the heat they can gain during the day to help fuel their activities after dark. Climbing geckos may raise their temperatures by sitting underneath bark or rock flakes that

Basking helps to raise the body temperature of many geckos but is most critical for those diurnal species for which sunlight is limited. Bradfield's Namib Day Gecko (*Rhoptropus bradfieldi*) lives on the coast where fog and cool winds keep air temperatures low for much of the time. Whenever the sun appears, they rush out of crevices to expose their dark bodies on boulder surfaces.

Courtesy of Randall D. Babb.

are in the sun, whereas burrowing species may move to the warmest parts of their burrows, usually near the surface. Like basking, thigmothermy can warm the gecko to above the ambient air temperature, but the small size of most geckos does not let them retain heat very long. Some nocturnal geckos will risk exposure to predators in order to bask. This happens most often if the gecko can bask from some partly protected site, like the edge of a crevice, or in relatively warm places where the gecko's body temperature before basking is already high enough to allow it run or fight if need be.

Which animals eat geckos?

Because geckos are mostly small and lack venom or poison, they are sought after as food by all sorts of predators. Snakes are a major source of gecko mortality. Like geckos, they may be diurnal or nocturnal and occur in all types of habitats. Their slender, elongate bodies make it possible for them to reach geckos in retreats that might be safe from other attackers. There are even snakes that specialize in eating gecko eggs. Birds are also important predators of geckos. Everything from shrikes—which impale geckos on the spines of plants to feed on them later—to ostriches and chickens to owls can and do eat them. The study of owl pellets reveals that geckos were a major part of their diet on Pacific islands before rats were introduced. Geckos are too small for many carnivorous mammals to bother with, but there are exceptions. For example, meerkats and other small terrestrial carnivores will regularly eat ground geckos, and primates often take arboreal species. Rats, which are introduced pests on many islands take a terrible toll on geckos, eating both adults and eggs. Rats are especially dangerous for larger geckos that are too small to fend off the rodents but too large to fit into narrow cracks that are safe from their predations. Carnivorous lizards, including other geckos, will happily eat geckos too, and even larger frogs will eat geckos if they catch them. Some geckos are small enough that they can be eaten by arthropods and even moderate-sized geckos are sometimes captured and devoured by spiders, scorpions, solfugids, and centipedes. Insects, especially mantises and ants, will kill and eat geckos too.

Do geckos get sick?

Most geckos probably fall prey to predators before they ever get sick, but some unlucky geckos do suffer from a variety of ailments. For example, geckos can get various forms of cancer and respiratory, skin, and bone diseases. They are susceptible to a diversity of viruses, bacteria, fungi, protozoan parasites, worms, and ticks. Parasitic infections are responsible for a broad spectrum of sicknesses. Blood parasites, such as malaria and trypanosomes, are quite common in geckos in the wild but rarely make their hosts ill. The same is true of the small red trombiculid mites found on most geckos in the wild; these can lead to local skin damage in heavy infestations but do not act as disease vectors. However, other protozoan parasites that can easily be spread, especially in crowded or unsanitary captive conditions, can cause serious health issues like cryptosporidiosis and coccidosis that can ultimately result in death. Nematode worms in the gut, pentastomids in the lungs, and tapeworms and others in the internal organs may be tolerated

Nocturnal predators of all kinds feed on geckos. In the arid Southwest of the United States, the Western Banded Gecko (*Coleonyx variegatus*) falls prey to an Elf Owl. Apparently tail autotomy was not a successful defense in this case. Courtesy of Randall D. Babb.

Predators, like the Cape Coral Snake (*Aspidelaps lubricus*), take a toll on geckos, which once discovered have few effective defenses against such large foes. Here the posterior portions of a Quartz Gecko (*Pachydactylus latirostris*) are all that remain visible. Courtesy of Bill Branch.

at low levels of infection but can kill if untreated. Perhaps the most common sickness of captive geckos is metabolic bone disease. This is caused by insufficient calcium or vitamin D in the diet and affects the skeleton and other tissues of the body. Other vitamin deficiencies, as well as an excess of calcium or certain dietary supplements, can also cause health problems. Intestinal blockages caused by the ingestion of sand or other foreign bodies, or even by the presence of too much chitin from insect exoskeletons, can cause obstructions that can ultimately lead to death, as can egg retention by females. These conditions must sometimes be resolved surgically. A gecko weakened by any type of sickness is also more likely to succumb to secondary infections, like bacterial pneumonia, that worsen the lizard's condition.

How can you tell that a gecko is sick?

Typical signs of a sick gecko are loss of appetite, regurgitation, diarrhea, and loss of body weight. Cryptosporidiosis affects the digestive tract, and the gecko stops feeding. Eventually this leads to the use and depletion of the fat reserves in the tail (known as "skinny tail" syndrome) and eventually to emaciation. Intestinal obstructions and egg retention are other causes of appetite loss and can be fatal. Lesions, cysts, and other abnormalities of the skin, as well as respiratory and digestive distress, can be signs of a diversity

Geckos: The Animal Answer Guide

Geckos are susceptible to a variety of health problems. This Tokay (*Gekko gecko*) is emaciated, probably as a result of cryptosporidiosis, a condition caused by a parasitic protozoan.
Courtesy of Tony Gamble.

of parasitic infections. In metabolic bone disease, fragile bones collapse or are easily damaged and result in kinking of the spine, rubbery legs, and softening of the skull. This in turn makes the gecko lame, may prevent feeding, can lead to swelling of the limbs, and can induce skin lesions. In some *Rhacodactylus* an aptly named and self-explanatory condition called floppy tail syndrome can occur. General lethargy or unexpected changes in behavior can be other signs of sickness. In the case of captive geckos, owners should be on the lookout for any of these symptoms and should seek veterinary advice early to prevent harm to their pet.

Chapter 6

Reproduction and Development

How do geckos reproduce?

Most geckos have both male and female individuals, although there are some with only females. In geckos with two sexes there may be some sort of courtship in which the geckos use various signals to identify each other as (1) members of the same species, (2) members of opposite sexes, and (3) receptive to mating. The initial signals are sent and received at some distance, by visual and auditory means. At closer range chemical signals are used and finally the geckos may touch and even lick each other. Male and female geckos usually do not differ much in size, but in some genera they may be differently colored. Typically the male approaches the female and, if she is receptive, he bites her on the neck and positions his vent or opening of the cloaca (the common opening into which waste products and gametes are released) under hers. As in all amniotes—animals with a terrestrially adapted egg, such as reptiles and mammals—fertilization is internal. Male geckos possess a pair of structures called hemipenes (singular: hemipenis) that are used to transfer sperm to the female. At rest, these are turned inside out and lie inside the base of the tail. When mating, one of the hemipenes (they generally alternate in use) is everted through the vent and inflated with blood. One side of the hemipenis has a sulcus, or groove, in it that carries the sperm. The rest of the structure is decorated with a complex and species-specific spines, lobes, and other surface features. Although not quite a "lock-and-key" mechanism, the specific structures of the hemipenis probably help to prevent mismatched matings between different species. Copulation in geckos usually lasts only a few minutes. In some geckos females can store sperm for a long period, and it is possible

As is typical in gecko copulation, the male of this Kandyan Gecko (*Cnemaspis kandiana*) bites the female on the nape and brings his cloaca under hers in order to transfer sperm through one of his two hemipenes. Courtesy of Anslem de Silva.

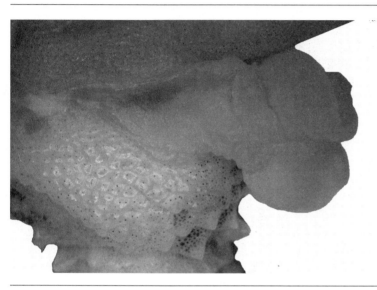

The everted hemipenis of the Madagascan Ground Gecko (*Paroedura picta*) showing its relationship to the tail base and the cloaca. Only one hemipenis, the left or the right, is used at a time. Courtesy of Tony Gamble.

that she may have multiple matings. Male geckos may also mate with several different females.

How can you tell the sex of a gecko?

Male and female geckos usually look rather similar to each other, although in some species sexual differences in color patterns occur (see "Are male and female geckos differently colored?" in chapter 3). For these species, color alone allows males and females to be distinguished. In some geckos, males are larger than females, but in others the trend is reversed. Overall the differences in size are usually relatively small. Unlike some groups of lizards, like iguanas and chameleons, male geckos do not possess elaborate ornamentation of the body, such as brightly colored dewlaps, enlarged crests, or "horns" on the snout. In many species the only way to determine the sex of a gecko is to examine its cloacal region. The hemipenes of the males, when withdrawn into the body, produce bulges behind the animal's vent. Thus, the tail base, when viewed from below, is enlarged in males, but not in females. With gentle pressure on the tail base the hemipenes can be everted to confirm the sex, but this should only be done by experts, as it is also easy to hurt the gecko or cause it to lose its tail if done improperly. Males also have more well-developed cloacal spurs (also called "postcloacal tubercles"). These are raised scales or tubercles that project laterally or dorsolaterally from the sides of the tail base. Many geckos have precloacalor femoral pores, openings on the underside of the hindlimbs and groin through which a waxy glandular secretion is released. With a few exceptions, such pores occur only in males. However, females may have small dimples in the same position that can make sexing difficult, and many genera of geckos lack such pores altogether. Another way to check sex is to examine the abdomen for signs of eggs or enlarged follicles. As many geckos have weakly pigmented bellies, it is often possible to see these signs through the skin. Male geckos of some species often have a bigger head than do females, presumably due to larger jaw muscles used for biting rivals in territorial fights.

Do all geckos lay eggs?

No, there are a few geckos that give birth to live young. This switch from oviparity (egg-laying) to viviparity (live birth) has taken place only twice in gekkotan evolution, both in the Diplodactylidae: once in the lineage leading to the New Zealand geckos and once within the lineage of New Caledonian Giant Geckos. In both cases, courtship and mating are

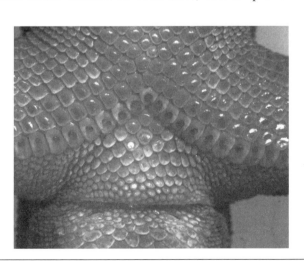

Precloacal and femoral glands are used by male geckos to mark their territories or make their presence known, chemically, to others of their species. In this Gold Dust Day Gecko (*Phelsuma laticauda*), the waxy secretion plugs can be seen in the chevron of scales anterior to the vent. Courtesy of Tony Gamble.

similar to that of other geckos, but, instead of the female laying down an eggshell over the fertilized egg in the oviduct, only the fetal membranes are present. The developing embryo is retained in the oviduct, where the dense capillary network of the fetal membranes are in close contact with the blood vessels of the mother. Although there is no direct connection between the two, their close association allows for gas exchange to take place. The embryo's nutrition, however, comes from the yolk that was present in the egg at the time of fertilization.

Because development takes place in the female's body, viviparity limits the number of times per year that geckos can reproduce. In the temperate to subtropical climates where the live-bearing geckos live, only one cycle of reproduction can usually take place in a year. In most viviparous species, two young are born after 3 to 5 months' gestation. However, the young of *Woodworthia maculata* in southern New Zealand, which are fully developed after 7 months, are retained overwinter to be born during the warm spring, after a total of 14 months. It has been argued that the New Zealand geckos live in one of the coolest environments of any gecko and that viviparity

All New Zealand geckos, like the Green Gecko (*Naultinus elegans*) shown here, as well as two species from New Caledonia, are viviparous and give birth to twins. This trait has only evolved within the Family Diplodactylidae. Courtesy of Bruce Thomas.

may be an adaptation to allow the female to keep her developing young warm by basking or otherwise directly controlling their thermal environment. Indeed, gravid live-bearing geckos are known to maintain higher body temperatures are than non-gravid females or males. This may be an advantage for geckos in harsh environments, but it does not explain why some, but not all, of the New Caledonian geckos also use this strategy. In any case, because they have such a limited reproductive output, viviparous geckos need to breed over many seasons to ensure lifetime reproductive success, and most, therefore, have long life spans by gecko standards.

Although viviparity may increase the chances of the embryo's survival and rapid growth, it also has some high risks for the mother. For many months, she must carry the additional weight of the developing young. This almost certainly places her at a disadvantage with respect to maneuverability and speed and at greater predation risk, particularly as she may need to expose herself more than a non-gravid female would to raise her body temperature.

The limitation of live birth to just one gecko family may have to do with the nature of gecko eggshells. Only pygopodoids (including the Diplodactylidae) and the Eublepharidae retain the primitive squamate condition of flexible, leathery-shelled eggs. In the other gekkotans, the eggs have a rigid calcareous shell. Both physiologically and evolutionarily, it is almost certainly easier to make the transition to not having a shell (and having a live birth) from the leathery eggshell than from the evolutionarily more recent rigid-shelled egg.

How long do gecko eggs take to hatch?

Aside from the viviparous geckos, geckos do not retain eggs in their bodies for long periods. However, geckos with leathery-shelled eggs do keep the eggs in the oviduct for several days longer than do geckos with rigid eggshells. This may be because it is more difficult for embryos of the latter to get sufficient oxygen in the oviduct. Embryos in newly laid eggs of eublepharid geckos, for example, are several stages more advanced than those in comparable gekkonid eggs. Depending on the temperature at which they are incubated, gecko eggs normally hatch in 40 to 110 days, most often 60 to 80 days for eggs at a temperature of 77°F (25°C). In captivity, with artificial heat sources, hatching can take as little as 33 days. There are exceptions, however. *Christinus guentheri* eggs from Lord Howe Island took 210 to 273 days to hatch. At the extreme, eggs of *Colopus kochi*, a gecko from the gravel plains of the Namib Desert, take more than 480 days to hatch and may sometimes require more than 600. Such a long period of development suggests that the embryos undergo what is known as a "diapause," a period when growth is halted and the embryo remains at the same stage of development for a long time. Although this is known in some other lizards, it seems to be very rare in geckos. The cool, foggy conditions of the habitat where this gecko lives are undoubtedly related to this long incubation period.

Where do geckos lay their eggs?

The leathery-shelled eggs of eublepharid and pygopodoid geckos may be laid under stones, bark, or logs or in a shallow pit excavated by the mother. Many climbing geckos will descend to the ground to lay their eggs at the base of trees or bushes under debris or in the soil where conditions of temperature and moisture may be favorable. Usually such nests are some distance from the mother's own normal retreat, presumably so that her activities do not bring attention to the location of the nest. Eggs in ground nests are sometimes laid communally and may even contain the eggs of several species of geckos. Communal nests may reflect that there are a limited number of optimal laying sites.

Geckos that lay rigid-shelled eggs can be either "gluers" or "non-gluers," with adhesive and non-adhesive eggs, respectively. Non-gluers lay their eggs in ways similar to geckos with leathery-shelled eggs. Gluers, on the other hand, stick their eggs to a surface such as a wall, rock, or plant leaf. To do this females carefully manipulate the eggs with their hindfeet and place them in a particular spot where their sticky coating causes them to adhere. Typically the eggs (if the clutch size is two) of gluers are placed next to each other and harden together as they dry. The shapes of glued eggs are always somewhat modified by their placement—at least the side of the egg that is attached to the substrate is partly flattened. In Namib Day Geckos, *Rhoptropus*, the eggs are broadly attached to stones and take on a hemispherical or even more flattened shape. Many egg-gluers are arboreal

Koch's Ground Geckos (*Colopus kochii*) lay eggs that undergo a diapause in development and can take more than a year and a half to hatch. Most gecko eggs hatch within 3 months, but the exact timing is largely dependent on their incubation temperature. Courtesy of Randall D. Babb.

Reproduction and Development

or rupicolous (rock living) climbing geckos. By gluing their eggs they can carefully chose egg-laying sites that will be largely inaccessible to predators. Such eggs are often found attached to ceilings or rocky overhangs or deep inside rock cracks. Such ideal spots for eggs may be quite limited, and it is relatively common for egg-gluers to lay communally. Such sites are usually favorable year after year so they are used again and again, sometimes by large numbers of females. Such sites often have viable eggs along with "egg scars," the remnants of the glued parts of eggs that have hatched in previous years. In extreme cases, populations of geckos may use the same sites continuously for so long that many layers of eggs accumulate on top of one another. Good examples of this can be seen in *Ptyodactylus* spp. and *Calodactylodes* spp. In the former genus I have seen entire ceilings of unoccupied buildings covered by generation upon generation of egg scars, suggesting that the site had been used for dozens if not hundreds of years.

Why do geckos lay hard-shelled eggs?

Nearly all lizards and snakes that lay eggs produce leathery-shelled (or parchment-shelled) eggs that have a mostly fibrous construction with a thin layer of overlying calcium crystals. Three families of geckos, however, Gekkonidae, Sphaerodactylidae and Phyllodactylidae, lay eggs that have rigid, brittle, calcareously shelled eggs that seem more like those of birds than other squamates, with the mineral layer, rather than the fibrous layer predominating. Although they are soft and pliable when laid, the eggs harden within a few hours of deposition, sometimes only in minutes.

Egg-gluing geckos are able to place their eggs in areas that may be difficult for predators to reach. The Sri Lankan Golden Gecko (*Calodactylodes illingworthorum*) lays its eggs communally here, for example, on a vertical cave wall. In addition to new eggs, the egg scars of the clutches of previous years are visible.

Geckos: The Animal Answer Guide

They are usually more spherical and less elongate than leathery-shelled eggs. These hard-shelled eggs have a very low rate of water loss compared with leathery eggs, which normally must take in some water from the environment during incubation and are ordinarily laid in protected sites with high humidity. In contrast, hard-shelled eggs do not require outside water to develop and can be laid in almost any environment, including very dry or exposed places. This may explain, in part, why the three families that have this egg type account for 85 percent of all gecko species and why only hard-shelled geckos have been successful on all continents. The hard shell also provides some level of resistance to exposure to seawater. Experiments have shown that gecko eggs can survive in seawater long enough to allow them to be transported by ocean currents. It is this capability that certainly accounts for the presence of geckos on even the most remote of islands in the Pacific Ocean. Eggs laid on bark or foliage are periodically naturally dispersed in mats of floating debris in storms, and those laid in ships or in cargo are carried around the world by humans. Not surprisingly, all of the species of geckos that are invasive or that have transoceanic distributions also have hard-shelled eggs.

Despite the advantages of hard shelled-eggs, there are also costs. Although less permeable to water than leathery-shelled eggs are, they are also less permeable to gases, limiting the absolute size of the eggs because of constraints on the diffusion of oxygen (bigger eggs have greater volume and therefore greater oxygen demand and at some point the surface area of the egg membranes—where gas is exchanged—can no longer meet this demand; because of the lower gas permeability of hard-shelled eggs, this

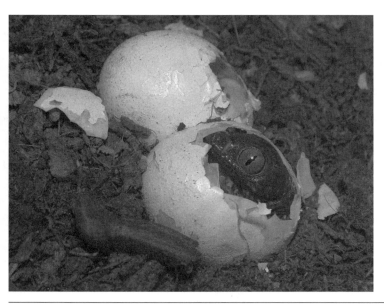

Rigid-shelled (calcareous) eggs like the one from which a tiny Tokay (*Gekko gecko*) is emerging, are typical of the families Sphaerodactylidae, Phyllodactylidae, and Gekkonidae. Although they place some physiological constraints on the developing embryos, these eggs have also allowed geckos in these groups to colonize over water and probably contribute to the species-richness of these families. Courtesy of Tony Gamble.

limit is reached at a smaller egg size). It also leads to more spherical eggs, as this shape maximizes the surface area-to-volume ratio and thus the effectiveness of gas exchange. The same ratio also minimizes the amount of calcium needed to build the shell, which could be important if this is in any way a limited resource. Egg size is also constrained because the size of the female's pelvic opening limits the dimensions of the rigid egg that must pass through it. Because hard-shelled eggs must contain all the water the embryo needs for development, there is less room for yolk, and, with this reduction in energy, the size of the hatchling that can be produced is limited. However, leathery-shelled eggs take up water after being laid. The egg, as it is laid, contains a higher proportion of energy stores. In addition, because they are more elongate and not limited by the need to maximize surface area, a greater volume of yolk, and thus ultimately a bigger hatchling can result from a leathery-shelled egg compared with a rigid-shelled egg of the same diameter. All of these factors mean that geckos with rigid-shelled eggs tend to produce both absolutely and relatively smaller eggs, with less energy input into each egg, than do other geckos. Smaller eggs yield smaller hatchlings, and these generally produce smaller adults. Thus, among egg-laying geckos, the very largest geckos are species with leathery-shelled eggs (*Rhacodactylus leachianus*) and, on average, geckos in the families with such eggs are larger than those in the hard-shelled families.

Do geckos only lay eggs once a year?

Eggs are costly for female geckos to produce, so it is only possible for them to lay a clutch once they have stockpiled sufficient nutrients to convert them into yolk to supply the developing embryo with food. As a consequence, most geckos that have a shortened activity season due to cool winter temperatures, or those that are active but resource limited (like some geckos in areas with distinct wet and dry seasons), produce only one clutch of eggs per year, usually during late spring or early summer, or during the wetter part of the year. Live-bearing geckos give birth at most once a year because they experience a long cool season and because the developing young occupy the female's reproductive tract for a minimum of 3 months. Geckos that are active and can feed year-round are likely to produce multiple clutches. With tropical geckos, it is not uncommon for five to seven or even more clutches to be produced per year. In captivity some geckos can produce more than a dozen clutches of eggs per year, but this is certainly more frequent than reproduction in the wild ever is. Species with multiple clutches usually space them out by periods of 3 to 5 weeks, but this varies considerably and in captivity there can be as little as two weeks between

Burton's Snake Lizard (Pygopodidae: *Lialis burtonis*)**; Australia.** Courtesy of Tony Gamble.

Pink-tailed Worm Lizard Pygopodidae (*Aprasia prapulchella***); Southeastern Australia.** Courtesy of Tony Gamble.

Amaral's Naked-toed Gecko (Phyllodactylidae: *Gymnodactylus amarali*); Brazil. Courtesy of Tony Gamble.

Midline Knob-tailed Gecko (Carpho-dactylidae: *Nephrurus vertebralis*); South and Western Australia. Courtesy of Brad Maryan.

Muller's Velvet Gecko (Gekkonidae:
Homopholis mulleri)**; Northeastern
South Africa.** Courtesy of Johan Marais.

**Big-scaled Dwarf Gecko (Sphaero-
dactylidae:** *Sphaerodactylus macrolepis grandisquamis*)**; Puerto Rico.**
Courtesy of Tony Gamble.

Tokay Gecko (Gekkonidae: *Gekko gecko*); Tropical East and Southeast Asia. Courtesy of L. Lee Grismer.

Giant Madagascan Leaf-tailed Gecko (Gekkonidae: *Uroplatus giganteus*); Madagascar. Courtesy of Miguel Vences.

Gold Dust Day Gecko (Gekkonidae: *Phelsumala ticauda*); Madagascar.
Courtesy of Tony Gamble.

Rough Thick-toed Gecko (Gekkonidae: *Pachydactylus rugosus*); South-western Africa. Courtesy of Bill Branch.

Madagascan Big-eyed Gecko (Gekkonidae: *Paroedura masobe*); Madagascar. Courtesy of Tony Gamble.

Brazilian Gecko (Phyllodactylidae: *Phyllopezus pollicaris*); Southern Brazil and adjacent areas. Courtesy of Tony Gamble.

African Wall Gecko (Phyllodactylidae: *Tarentola ephippiata*); North Africa. Courtesy of Laurent Chirio.

Large-scaled New Caledonian Chameleon Gecko (Diplodactylidae: *Eurydactylodes symmetricus*); New Caledonia. © Tony Whitaker.

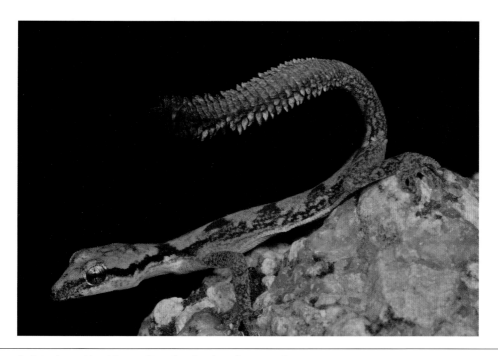

Feather-tailed Gecko Gekkonidae: (*Afrogecko plumicaudus*); Angola. Courtesy of Bill Branch.

Barking Gecko (Gekkonidae: *Ptenopus garrulus*); Southern Africa.

Courtesy of Johan Marais.

Soft Spiny-tailed Gecko (Diplodactylidae: *Strophurus spinigerus inornatus*); Western Australia. Courtesy of Brad Maryan.

clutches. In arid environments, as in parts of Australia, gekkonoid geckos may have an advantage over other lizards, as their hard-shelled eggs allow them to lay eggs during drier periods of the year when the eggs of leathery-shelled species would dehydrate.

Females of many geckos are capable of sperm storage, so they can mate once and keep the sperm alive in their bodies for later use for many months. Thus, females that lay many clutches of eggs in a year may not have to mate many times but rather can use sperm from an earlier mating to fertilize their eggs. This capability is another reason why some geckos are good dispersers and colonizers: only one female with stored sperm is necessary to found a new population. Sperm storage also allows geckos in environments with short active seasons to mate in the autumn and fertilize their eggs as early as possible in the next spring, allowing an earlier hatching and affording their young the maximum opportunity to feed and grow before the onset of the next winter. *Homonota darwini*, the southernmost gecko in the world, has a one egg clutch and only reproduces once per year to once every 2 years, giving it one of the lowest annual reproductive outputs of any lizard.

How many babies do geckos have?

Virtually all geckos that lay eggs produce either one or two eggs per clutch. Those that are viviparous always have twins. Two eggs are common in all pygopodoids and in eublepharids. Among the other families, a single egg is seen in representatives of numerous miniaturized lineages, such as the New World sphaerodactyls. One-egg clutches are also seen in some larger-bodied sphaerodactylids, like the West Indian *Aristelliger*, raising the question of whether these are derived from small-bodied ancestors or whether something other than size is related to a reduction in clutch size. Some other geckos that produce only one egg are the large neotropical phyllodactylid *Thecadactylus rapicauda* and the African gekkonid *Ptenopus garrulus*. For the most part, all species in a genus lay the same number of eggs, but in *Gehyra* some species have two egg clutches and others one. The same is true for two subspecies of the phyllodactylid *Gymnodactylus geckoides*. Geckos that produce only one egg often produce larger eggs with more energy stores than their relatives that have two-egg clutches. One measure of the reproductive effort of a gecko is its relative clutch mass (RCM). This is the ratio of the mass of the eggs to that of the female after she has laid them. In comparison with other lizards, geckos, despite their small clutches of one or two eggs, have more-or-less average RCMs. RCM is often high in lizards that are ambush predators and do not need to move

Most geckos lay clutches of two eggs, like the Quartz Gecko (*Pachydactylus latirostris*), but the number of clutches per year can vary a great deal, with cool temperate species breeding once a year or once every other year and tropical species having many clutches. Courtesy of Jon Boone.

a lot and small in active foragers that move constantly while hunting. Not surprisingly, geckos, which are intermediate in their foraging mode, are also intermediate in their RCM.

A few species of gecko have been reported to produce three-egg clutches, but some of these records may reflect instances of communal nesting by two geckos. Because geckos are constrained to produce only very small clutches, their success depends either on ensuring high survival rates for their young or on being able to produce many clutches in a year (or in a lifetime). Thus, the number of eggs can be related to eggsize, juvenile growth rate, and longevity.

How is the sex of a baby gecko determined?

In most geckos the sex of the offspring is determined genetically, as it is in humans. although in geckos usually the male has two copies of the same type of sex chromosome, whereas the female is heterogametic (having two different copies of the sex chromosome, one that will produce male offspring and the other female)— the reverse of humans. The Tokay Gecko, *Gekko gecko*, has been reported to have male heterogamety, although other members of the genus show female heterogamety and still others may have temperature-dependent sex determination (TSD), in which the temperature experienced by the developing embryo during a critical window of development (typically the first 3 weeks) determines what the sex of the gecko will be. In Leopard Geckos, females are produced at both low (79°F, or 26°C) and high (93° F, or 34°C) temperatures, with males predominating at warmer intermediate temperatures. Most eublepharid geckos seem to have TSD, although in *Coleonyx* sex is genetically determined. TSD is also known in some species of New Caledonian diplodactylids, including the commonly kept Crested Gecko, *Correlophus ciliatus*, and the Giant Gecko *Rhacodactylus leachianus*. In these species more females are produced at low temperatures and more males at high temperatures. TSD is known to occur in some *Phelsuma* species, but the sex-determination mechanism for most *Phelsuma* remains unknown. All other gekkonids that have genetically determined sex and for which there are data have female heterogamety. Male heterogamety is known from pygopods and sphaerodactylids. At least some phyllodactylids have TSD, but so few species have been investigated that it is impossible to make broad generalizations. Identifying the mechanism of sex determination requires the study of karyotypes (the number and structure of chromosomes), or carefully documented breeding experiments with the temperature strictly controlled.

Are some gecko species all female?

At least eight species (or species complexes) of geckos are parthenogenetic. That is, they are all-female species that reproduce without the need for males, producing female offspring genetically identical to themselves. Such geckos originate from the hybridization of two different parental species. Parthenogenetic geckos have an advantage over sexually reproducing geckos under certain circumstances. Only one individual is necessary to colonize a new area, such as an isolated island or a region recently disturbed by fire, and once established unisexual lizards are able to increase their population size at twice the rate of bisexual species, in which only half the population (the females) actually lay eggs. However, because they do not undergo genetic recombination as sexual species do, they have very limited variation and may not have the genetic diversity to respond to changes in

Most eublepharid geckos, including the Vietnamese Leopard Gecko (*Goniurosaurus araneus*), exhibit temperature-dependent sex determination, as do a variety of other geckos. However, most geckos do appear to have genetic sex determination. Courtesy of Tony Gamble.

Reproduction and Development

The Mourning Gecko (*Lepidodactylus lugubris*) is one of a small number of geckos that have all female populations and reproduce by cloning themselves. Because no males are needed, a single female can found a new population. Not surprisingly, such parthenogenetic geckos are often the only lizards on remote islands. Courtesy of L. Lee Grismer.

the environment. Not surprisingly, several of the most widespread species of geckos are parthenogens, these include the Mourning Gecko (*Lepidodactylus lugubris*), *Hemidactylus garnotii*, *Hemiphyllodactylus typus*, *Nactus pelagicus*, and *Heteronotia binoei*.

Study of parthenogenetic species has revealed that there are multiple different "chromosome races" in many, and in some cases the animals are diploid and in other triploid. This demonstrates that similar parthenogenetic lineages have formed on multiple occasions from different hybridization events. Calling all such chromosomally distinct lineages the same species is certainly underestimating the diversity of these lineages, but traditional views of species and the Linnaean system of binomial nomenclature were not developed with such animals in mind. Indeed, it was not until the late 1960s that unisexuality was confirmed in geckos.

There is evidence that other geckos, most notably the Giant New Caledonian Gecko *Rhacodactylus leachianus*, which normally reproduce with both male and female genetic contributions, can, under certain captive circumstances produce viable offspring without the participation of a male. The mechanism by which this occurs, the trigger for it to happen, and whether this occurs in nature remain unclear, but a similar phenomenon takes place and is more well documented in a number of snake species.

Do geckos care for their young?

Most geckos do not exhibit any form of parental care beyond the careful concealment of eggs or the selection of eggs sites favorable to incubation. Once laid, eggs are abandoned. A possible exception occurs in viviparous geckos. Because of their tremendous physiological investment in their young, as well as the limitation placed by long gestation on total number of young produced in a lifetime, it make sense that live-bearing geckos would be those most likely to demonstrate some sort of care. Minimally, in those viviparous species studied, adults are tolerant of the presence of the offspring for relatively long periods of time. Thus, indirectly, the young may gain some protection from the proximity of the larger adults. Mothers of the New Caledonian Rough-snouted Gecko *Rhacodactylus trachyrhynchus* supposedly provide some care to their babies after they are born, at least in captivity.

How fast do geckos grow?

Geckos grow quickly during the first weeks and months of life. This can be seen by the high frequency of skin shedding in healthy young geckos. The rate of growth is influenced by the temperatures the geckos experience and their nutritional state. In most smaller geckos, maturity is reached in from one to three seasons in the wild, but the parthenogenetic *Lepidodactylus lugubris* can reproduce at as little as 8.5 months. In some larger geckos, and particularly in larger geckos in cooler climates, it may take as much as 7 years for geckos to become reproductively active. In the temperate Patagonian species *Homonota darwini* males take 9 years to reach maturity, whereas females take 5. In captivity, these times may be much shorter, as higher temperatures and higher rates of food consumption allow geckos to reach reproductive sizes earlier. Female geckos are usually about 70–80 percent of their maximum size when they reach maturity, but males are somewhat smaller, probably because they do not make such a big energy investment in reproduction. For females, reproduction comes at the expense of growth. For males mating probably does not delay growth significantly. Once geckos reach maturity, growth slows but may never totally cease. At least in Leopard Geckos, growth rate later in life may be affected by the incubation temperature experienced by the eggs.

How long do geckos live?

Lifespan in geckos depends on several factors. One is body size. Smaller geckos generally have shorter lifespans than do large species. So do species living in warmer conditions. There is also an evolutionary component, as members of some gecko families generally live longer than those of others. Pygopodoids, in particular, have the longest documented life spans (Table 6.1). Not surprisingly, the longest lived gecko thus far documented is a large-bodied live-bearing species, *Hoplodactylus duvaucelii* from New Zealand. Like other live-bearing geckos, it can only reproduce once per year or

Table 6.1. Maximum known longevity for selected species of geckos.
As a rule, larger geckos live longer than smaller species and members of certain families (Diplodactylidae, Eublepharidae) tend to live longer than others. Accurate ages are not available for most species of geckos in the wild, and captive longevity can be difficult to interpret. Some species live much longer than in the wild because they are safe from predation and receive adequate food, but in other cases captive life span may be less than in the wild because their husbandry is not understood.

Species	Maximum age (years)	Captive or wild
Woodworthia maculata	37	Wild
Hoplodactylus duvaucelii	36	Wild
Coleonyx variegatus	34.8	Captive
Rhacodactylus leachianus	30	Captive
Eublepharis macularius	29	Captive
Naultinus grayi	23	Captive
Euleptes europaea	21	Captive
Gekko gecko	20	Captive
Naultinus elegans	20	Captive
Homonota darwini	17	Wild
Hemitheconyx caudicinctus	16.2	Captive
Toropuku stephensi	16	Wild
Tarentola mauritanica	14	Captive
Oedura monilis	14	Captive
Heteronotia binoei	13.6	Captive
Phelsuma guentheri	13.5	Captive
Gonatodes albogularis	13	Captive
Ptyodactylus hasselquistii	12.6	Captive
Teratoscincus scincus	12.5	Captive
Hesperoedura reticulata	11	Wild
Gehyra variegata	10	Wild
Geckolepis typica	10	Captive
Uvidicolus sphyrurus	10	Captive
Stenodactylus doriae	8.8	Captive
Hemidactylus turcicus	8	Captive
Phelsuma guimbeaui	8	Captive
Uroplatus sikorae	7	Captive
Rhoptropus diporus	6	Captive
Paroedura picta	5	Captive
Strophurus elderi	4.7	Captive
Mediodactylus kotschyi	4	Captive

Larger geckos tend to live longer than smaller geckos, and pygopodoid geckos live longer than their gekkonoid relatives. Duvaucel's Gecko (*Hoplodactylus duvaucelii*), the largest living New Zealand gecko, is known to live at least 36 years in the wild. © Tony Whitaker.

even once every other year. Thus, it must live a long time in order to ensure that it produces offspring that will themselves live to reproduce. Studies suggest that this species is not sexually mature until about the age of 7, and it has been confirmed to live a minimum of 36 years in the wild. Captive specimens of other viviparous geckos have lived at least 37 years. The best records for gecko longevity come from captive geckos, and these are probably not typical of wild individuals, which certainly suffer high rates of mortality from predators and other sources. Captive Leopard Geckos, *Eublepharis macularius*, have lived for at least 29 years, and *Rhacodactylus* species can certainly live for well over 20 years, more than 30 in the largest species. *Homonota darwini*, which takes a long time to reach maturity and has a very low reproductive rate has a relatively long lifespan of 17 years for a small gecko. It is probable that most small- to average-sized geckos in the wild live only 2–8 years.

Reproduction and Development

Chapter 7

Foods and Feeding

What do geckos eat?

Nearly all geckos will eat almost any type of arthropod that they can overpower and ingest. Very tiny geckos, like the Dwarf Geckos *Coleodactylus* or *Sphaerodactylus* may eat springtails (collembolans) and mites, whereas large geckos may eat larger prey such as orthopterans, beetles, and scorpions. Although some geckos specialize on certain types of prey, most feed on a wide variety, reflecting the relative local diversity and abundance of potential food organisms. Some geckos capitalize on prey items that are superabundant. Barking Geckos, *Ptenopus garrulus*, collected in Namibia on a rainy night when winged termites were flying had, in some cases, eaten more than half of their own body weight in these insects! The Australian geckos *Diplodactylus conspicillatus* and *Rhynchoedura ornata* are among a small set of arid zone geckos only eat termites. When they locate a termite mound or trail, they can fill their stomachs. Termites are patchily distributed, however, and geckos specializing in eating them must endure lean periods in between the good times.

Geckos cannot digest leaves and stems easily and only eat them accidentally if ingested with animal prey. However, many geckos will eat high energy, easily digestible plant material, like fruit, sap, nectar, or pollen. This is especially true of geckos occurring on islands, like New Zealand, New Caledonia, Madagascar, and the Mascarene Islands. In some of these areas geckos may take the place of insects as important pollinators and seed dispersers. Larger geckos will eat other vertebrates if they can subdue them. For most geckos, only other small lizards, sometimes of their own species, are small enough. However, the very largest geckos have broader

Insects and other arthropods make up the bulk of the diet for most geckos, including the Leopard Gecko (*Eublepharis macularis*), making them one of the most important of all nocturnal insectivorous vertebrates after bats. Courtesy of Tony Gamble.

Island geckos often eat fruit, nectar, or pollen and may be important pollinators or seed dispersers. The Bluetail Day Gecko (*Phelsuma cepediana*) of Mauritius is the only known pollinator and seed disperser of the rare plant *Roussea simplex*. Courtesy of D. M. Hansen.

diets. *Rhacodactylus leachianus* will take nestling birds, and *Gekko gecko* has been observed to eat snakes and small mammals. The Niah Cave Gecko of Borneo, *Cyrtodactylus cavernicolus*, takes advantage of the swiftlets that nest in the caves where they live and eats hatchling birds that occasionally fall from the cave walls. The only geckos known to truly specialize on vertebrate prey are the large pygopods of the genus *Lialis*. They are predators of skinks, which they grasp with their highly specialized jaws and teeth and swallow headfirst. Perhaps the broadest diet of any gecko is that of the Gargoyle Gecko, *Rhacodactylus auriculatus*. Although also a generalist insectivore, it is known to also eat snails, flowers, sap, skinks, and other geckos. Regardless of their diet, geckos, and especially nocturnal geckos, appear to have empty stomachs more often than other lizards do. This is particularly true at cooler times of year when both prey availability and the metabolic demands of the gecko are lower.

Foods and Feeding

How many teeth do geckos have?

Gecko teeth are small and numerous and usually similar in size and shape throughout the jaws. The teeth are present on the premaxilla and the maxillary bones in the upper jaw and on the dentary bone in the lower jaw. There are usually 9–13 teeth (typically a set number for a given species) in the premaxilla no matter how big the gecko is, but the number on the other elements is variable and increases as the animal grows and the dental lamina, the area where teeth form, elongates. The total of the teeth in the upper jaws is usually slightly greater than those in the lower jaw. The total number of teeth varies over time, as the teeth may be lost and replaced; thus, it is best to refer to the number of tooth loci, or the number of spots where teeth are either present or may be replaced. Most adult geckos have between 50 and 80 tooth loci in both the upper and lower jaws. However, the Leaf-tailed Gecko, *Uroplatus fimbriatus*, has the greatest number of teeth (more than 300 in total) of any gecko. This is probably the greatest number of teeth in any living species of terrestrial vertebrate. The reason for having so many teeth is unknown, as they have a fairly typical gecko diet. Most geckos have simple conical teeth, but the eyelid geckos have more complex cusps or ridges on the teeth. The Gargoyle Gecko, *Rhacodactylus auriculatus*, has relatively long, blade-like "fangs" that may help it puncture the skin of other lizards. In the skink-feeding pygopod *Lialis*, the teeth are directed posteriorly and are hinged to allow them to bend backward as the prey is swallowed.

Do geckos chew their food?

Geckos do not chew their food. Like other non-mammalian vertebrates, geckos' jaw muscles are largely arranged to permit up-and-down motion of the lower jaw and some anterior-posterior motion, but little to no side-to-side (propalinal) motion. Because geckos do not chew, there is not an exact alignment (occlusion) of the upper and lower teeth to slice, shear, or crush their food. Instead geckos bite down on their prey, piercing its body wall and usually killing or incapacitating it before using the tongue and movements of the head to force the prey backward so it can be swallowed. Prey may also be banged on the ground or other surfaces while it is being bitten. Insect or spider legs may sometimes be knocked off in the process, but geckos are capable of swallowing even larger, vertebrate prey intact.

Why do geckos' eyes sink in when they bite?

The eyes of geckos often sink down into their orbits slightly when they sleep. A permanent condition of deeply sunken eyes is a sign of poor health

Large-bodied Madagascan Leaf-tailed Geckos, like *Uroplatus henkeli*, have the same tiny, peg-like teeth that most geckos have, but, because of their large heads and jaws they may have more than 300 teeth, double what many geckos have and probably more than any other living amniote (reptiles, birds, and mammals). Courtesy of Tony Gamble.

and is typical of malnourished geckos. However, the eyes also "sink in" or are retracted when geckos bite. If you observe a gecko closely as it bites something, let's say your finger, you will notice that its eyes sink in or are pulled downward and backward. This action is the result of the retractor bulbi muscles that originate on the base of the skull and attach to the back of the eyeballs. This action occurs at the same time as the lower jaw is closed, the upper jaw is flexed downward and backward, and the bite force being generated is greatest. When feeding on large or potentially dangerous prey, or defending themselves against potential predators, the retraction of the eyes removes these large and delicate organs from the most direct sources of damage (things like the flailing legs of a grasshopper or the biting jaws of another gecko). Each eye can also be retracted independent of biting in response to a touch or even the threat of a touch. While the eyes are retracted the gecko's vision is obscured, so this is not a safe condition to maintain for an extended period. Thus, when the gecko stops biting down and its jaws are opened again, the eyes return to their normal position.

How do geckos find food?

Geckos have good senses and probably can and do use all of them in locating food. They have good vision.In nocturnal species color vision is possible even under extremely low light conditions. They are able to detect movement and identify types of prey even on dim moonlit nights. Geckos have two chemical senses that work over distance. One is smell, the other is vomerolfaction. Vomerolfaction is like smell, but it uses a special organ in the roof of the mouth (Jacobsen's organ, or the vomeronasal organ) to detect

Burton's Snake Lizard (*Lialis burtonis*) is the longest gekkotan lizard and one of few to specialize on vertebrate prey. Like other geckos, when it bites down, its kinetic skull allows its snout to flex and its eyes are partially withdrawn into their sockets. Courtesy of Tony Gamble.

particular types of chemical cues (see "How do geckos communicate?" in chapter 4). Snakes and lizards with deeply forked tongues, like monitors, have an especially well-developed vomeronasal sense, whereas geckos are believed to rely more heavily on smell or olfaction. These forked tongues can carry chemical cues directly into the openings into the vomeronasal organ and allow for directional discrimination (a stronger signal on one side indicates the direction of the source of the stimulus). Geckos have fleshy tongues with only a small notch at the tip, but still use the tongue to pick up chemicals to be delivered to the vomeronasal organ. Geckos are usually thought of as ambush or sit-and-wait predators because they spend a large part of their time motionless and will run short distances to catch passing insects. However, many geckos also actively hunt prey during bouts of activity during which they may be using a combination of visual and chemical cues. That both mechanisms are used is also shown by the fact that nocturnal geckos often eat both diurnal and nocturnal prey. The former are most likely located at night, when they are asleep or inactive, by smell or vomerolfaction, whereas the latter are probably ambushed. Geckos probably also use their excellent sense of hearing to locate prey, but this has not been studied.

Can geckos taste?

Some geckos have taste buds distributed over the surface of the tongue and parts of the lining of the mouth. However, geckos have been found to

With their excellent vision and hearing and good sense of smell, geckos use a diversity of means to locate their prey. They may either actively track down prey or, as with this Tropical House Gecko (*Hemidactylus mabouia*), wait for a victim to get close enough to be captured in a single lunge. Courtesy of Tony Gamble.

have the lowest number of taste buds on the tongue of any lizard group so far investigated. In some species, no taste buds have been found at all. In some geckos with taste buds on the tongue, an additional small set of taste buds is found on the roof of the mouth, next to the openings to the vomeronasal organ. This suggests that taste and vomerolfaction are related, just as smell and taste are connected in humans. Why taste buds are absent in some geckos remains unknown.

Are any geckos cannibals?

Like other lizards, many geckos will take advantage of smaller individuals of their own species as food sources. In the wild some geckos, such as the Indian *Hemidactylus prashadi*, may segregate microhabitat by age class in order to prevent juveniles from coming in contact with hungry adults until they reach larger sizes. Cannibalism in geckos is rarely documented in the wild, except for easily observed house geckos, but it undoubtedly occurs, with its likelihood increasing in times of food shortages and under crowded conditions. The cramped conditions of captivity increase the risk of cannibalism. It has been recorded in at least ten species of geckos, including some that are otherwise unknown to take vertebrate prey. With few exceptions, it is generally safest to keep hatchling and juvenile geckos separate from adults, as even parents will devour their own offspring.

Can geckos store energy from food?

Some geckos are quite efficient at storing energy. This is especially important in geckos that live in unpredictable environments where food availability may change over time. There may be long periods during which few prey are available or the gecko cannot forage. Energy is usually stored as fat in the tail. Geckos that rely on this source can have very thick or broad tails when they are well-fed. In some species, the tail may lose its normal smoothly tapering shape and develop a lobed appearance caused by the storage of a roll of fat within each segment. This does not occur in regenerated tails, which no longer maintain their original segmentation. Under certain conditions, geckos with extensive caudal fat stores can survive up to a year without food as long as they remain hydrated. Given the importance

Many geckos will eat smaller animals of their own species if given the chance. Prashad's Gecko (*Hemidactylus prashadi*) of India avoids this by having some spatial segregation of different age classes, which keeps juveniles and adults apart.

Courtesy of Tony Gamble.

Geckos store energy as lipids in their tails. A healthy and well-fed Yellow-bellied Gecko (*Hemidactylus flaviviridis*) exhibits distinct fat rolls at the base of the tail.

Geckos: The Animal Answer Guide

of the tail as a storage organ, it is not surprising that starving geckos are less likely to shed their tails than are well-fed ones. In some geckos, including *Rhoptropus* spp., fat is also stored in bands under the skin along the ventro-lateral margin of the body. Such alternative storage areas may be important for geckos that have specialized tail functions that would be incompatible with a fat tail.

Do geckos drink?

Yes, most geckos drink. Just like other animals, geckos need water to remain hydrated and in good health. In the wild, geckos get much of the water they need directly from the food they eat. Some geckos, particularly those from desert areas, rarely if ever need to supplement the dietary water they take in. However, even these geckos will take advantage of external sources of water if they are available. For most geckos, water is obtained by licking rain droplets off of surfaces, including leaves, stones, walls, and their own bodies. Some gecko species will drink from open water containers in captivity, but in the wild most geckos probably rarely encounter or use open water sources. Paradoxically, water availability is not a problem for geckos in the coastal Namib Desert, one of the driest places on earth. Cold ocean currents help to produce frequent foggy conditions during which water condenses directly onto the skin of geckos, where it can be lapped up. At least in some geckos the front part of the tongue is specially modified to uptake and channel water into the mouth.

Chapter 8

Geckos and Humans

Do geckos make good pets?

This depends on what one wants out of a pet. Geckos do not show affection to humans. Most gecko species do not like to be handled and may autotomize (or break off) the tail, bite, or defecate if they feel threatened or are treated roughly. Most are too small and delicate to be handled by small children, and those that are more robust are sometimes big enough to give a painful, though not dangerous, bite. In addition, most geckos do not like to be in the open, exposed to danger, and so prefer to hide in or under things and thus are not ideal pets to watch either. However, some geckos, most notably Leopard Geckos and Crested Geckos, are inexpensive and easy to care for and are reasonably tolerant of handling. For the right owner, someone who is committed to learning about and caring for the animal, geckos can be very rewarding to keep as pets. In addition to being beautiful, they have interesting behaviors. Their often slow and deliberate actions allow one to really appreciate their biology. Not all geckos are easy to keep, however, and gecko owners should be sure not to get in over their heads. Choosing a healthy, captive-bred gecko will maximize the owner's enjoyment and the gecko's quality of life and will not put a strain on wild populations.

How do you take care of a pet gecko?

There are many sources of information about keeping and breeding geckos in captivity, some treating the more commonly kept species in great detail. These should be consulted before anyone gets a gecko for a pet. To begin with, any new gecko should be inspected carefully to make sure that

Geckos can make excellent pets. Although they are neither affectionate nor obedient, some geckos, like Leopard Geckos (*Eublepharis macularius*) and a few others, are inexpensive, easy to care for, fascinating to watch, and can even tolerate gentle handling. Courtesy of Tony Gamble.

it is healthy. If a new gecko is added to an enclosure already housing one or more geckos, it should be quarantined first. Keeping the gecko on its own in a very basic cage for at least a month will allow one to most easily spot the symptoms of some of the diseases to which geckos are prone.

The right type of enclosure for the gecko must be provided. This does not mean that the terrarium must re-create the natural environment of the gecko, but it must meet certain requirements. First, it must be large enough to allow the animal to move around and to meet its needs. For terrestrial geckos, a large area of floor space is needed, whereas for climbers, a taller enclosure with three-dimensional space may be more appropriate. The enclosure should minimally allow space for one or more shelters or hiding places and a water dish (unless you will be misting (spraying water) for the gecko to lick up, in addition to some open space to allow the gecko to move and explore (and be observed). In some species, particularly terrestrial ones, geckos will use a defacatorium, a set place where it will always go to deposit its waste. This is usually a spot along a wall of the enclosure. The terrarium should be large enough to accommodate this at some distance from shelters and water sources. The choice of substrate for the gecko is also important. It should be something nontoxic, easy to clean, and comfortable for the gecko to walk on. Either artificial surfaces or those that are similar to the animal's natural substrate are appropriate. For climbing species, it is important that perches or surfaces of the correct

Geckos and Humans

Captive geckos need access to food, water, shelter, and a choice of temperatures. Sanitation and health concerns, as well as social interactions, should always be considered when multiple animals are kept in one enclosure. Under the right conditions, many geckos can be kept and cared for by a conscientious owner.

size and shape are provided. Rock dwellers should have rocks to climb on and arboreal species should have bark or some sort of vegetation (artificial or natural). Geckos are easily stressed if they do not have places to hide and rest that are protected, so these are especially important. Either artificial plastic hiding places or ones made of bark, hollow branches, or rocks can be used. These should be left undisturbed as much as possible. If a water dish is used, it should be shallow to prevent accidental drowning.

Another critical concern for keeping geckos is temperature. Ideally, geckos should be presented with a range of temperatures within their enclosure, allowing them to choose what they prefer. This can be provided by heating sources from above (lights or lamps) or below (heating pads or tapes) and should ensure that the gecko can access both cooler temperatures (room temperature) and warmer ones, but should not expose the gecko to dangerously high temperatures. Lighting is important for diurnal geckos, but less so for nocturnal ones that have other heat sources, although all geckos should experience periods of light and dark to promote normal activity cycles.

Nutrition is extremely important. Geckos must be provided with a diet that supplies them with all their requirements. Most geckos eat live arthropods, but commercially raised mealworms or crickets may not provide all of the necessary nutrients, so various types of supplementation are often needed. Diet should be tailored to the species of gecko being kept; larger

Geckos: The Animal Answer Guide

geckos will often accept "pinky" mice and some enjoy fruits or other sweet substances. It is especially important that geckos receive enough calcium and vitamin D to prevent the metabolic bone disease. Uneaten food should be removed from the cage to prevent insects from eating the gecko or from rotting.

Finally, it is important to know what other living things can safely be kept with a pet gecko. Toxic or irritating plants should be avoided and care must be taken if multiple geckos are kept together. Male geckos will usually fight or at least intimidate smaller males so it is not a good idea to keep them together, although a single male and female or a male and two females can usually be kept in harmony. Hatchlings and young should be kept in separate containers from adults to avoid cannibalism. These suggestions are all quite basic. Every species of gecko will have its own special needs. The real answer to "How do you take care of a pet gecko?" is this: prepare in advance, learn all you can, invest time into caring for your gecko, and turn to people with more experience when necessary.

Are geckos venomous?

No geckos are venomous and, aside from the slight mechanical damage done by the bite of some of the largest species, they are totally harmless. Among lizards, only Gila Monsters and Mexican Beaded Lizards (Family Helodermatidae), and possibly large monitors, such as Komodo Dragons (*Varanus komodoensis*), have venoms capable of affecting humans. Recent research suggests that venom is a primitive feature of a large group of lizards and snakes (the Toxicofera), even though in many members of this group, such as iguanid lizards, toxins dangerous to humans are not present and there is no specialized venom delivery mechanism. However, geckos are not members of the Toxicofera and do not appear to have had venom at any point in their evolutionary history. Nonetheless, people in many cultures have considered geckos to be venomous.

The Mediterranean gecko genus *Tarentola* gets its name from the town of Taranto, Italy, the same source as the word tarantula, originally used for the Wolf Spider, *Lycosa tarantula*. Both the gecko and the spider were once wrongly considered to be dangerously venomous. Early European reports of the Tokay Gecko also reveal that they were considered to be venomous by the Javanese. It was believed that the gecko could both kill with its bite and poison with its urine. In parts of North Africa and the Middle East, geckos were long believed to poison food merely by walking across it. Even today people in many cultures fear geckos and consider them to be venomous. I have experienced this in both southern Africa and in New Caledonia in the South Pacific. Since this belief is not based on fact but is nonetheless widespread across many cultures, it is likely that it derives from the common human tendency to ascribe sinister intentions and behaviors to things that are not understood. Especially to preindustrial societies living without artificial light, creatures of the night (bats, for example) were mysterious and ascribed deadly or supernatural powers. What could be more mysterious than a nocturnal gecko, hiding in the shadows with its large eyes and ghostly appearance?

Do geckos feel pain?

Yes, geckos feel pain. Pain is an adaptive response that protects animals from sources of potential harm. A stimulus that causes pain is likely to be harmful to the animal and so is avoided. A variety of receptor types in geckos, including specialized neurons in the skin, are capable of registering heat, pressure, or other stimuli that are painful above certain thresholds. Information from these receptors is sent to the spinal cord and may result in an immediate motor response, such as movement away from the painful stimulus, or may be relayed to the brain to be integrated with other sensory input (such as visual images) before a motor response is initiated. It is unknown how similar the sensation of pain in a gecko is to that of a human, but geckos respond predictably to painful stimuli. The most basic response is the rapid withdrawal of the body part or movement of the whole body away from the stimulus. In some cases, a distress vocalization may be made and, if escape from the source of pain is not possible, geckos will often bite down on the object, animate or not, that is causing the pain.

How can I see geckos in the wild?

Because of their generally small size and often nocturnal habits, geckos can be difficult to observe in the wild, especially without disturbing them. An exception are so-called house geckos. These are geckos of several different genera that commonly occur in or around human habitations, often to take advantage of the insect prey attracted by artificial lights. One or more house gecko species occur in most areas of the tropics and most of the alien geckos that have become established outside their natural ranges are among these. Such species can easily be observed at night around outdoor lights or on house walls or ceilings. Indeed, some of the most detailed studies of gecko behavior have been based on these species because of the ease with which they can be observed. The other geckos that may be easy to see in the right places are diurnal geckos. Day Geckos of the genus *Phelsuma* are often highly conspicuous. Some are large enough to make observations easy with binoculars. Nocturnal geckos in natural habitats are usually difficult or impossible to observe casually. Approaching geckos closely, especially with powerful headlamps or flashlights, will certainly disturb them and may also make them more vulnerable to predators so should be avoided. However, geckos may safely be observed from a distance using binoculars and a headlamp. With practice they can be located by their red eyeshine and their movements observed. For anyone with access to a night vision scope or goggles, this is another way to observe geckos without disturbing them.

Do people eat geckos?

Because they are relatively small, most geckos would probably not make a very good meal for a person. Also, because they are mostly generalist insectivores, the relatively little meat they have would probably not taste particularly good to the human palate. Indeed, most of the lizards that are eaten around the world are large and either herbivorous (like *Iguana*

Nocturnal geckos are difficult to observe without disturbing. Exceptions are geckos that are active in, on, or around human habitations, like these White-spotted Wall Geckos (*Tarentola annularis*),whose social interactions and predatory behavior can be observed on building walls. Courtesy of Laurent Chirio.

iguana) or carnivorous (*Varanus* spp.). However, there are exceptions. The large (5 inch/152 millimeter head and body length) *Gehyra vorax* was formerly eaten in Fiji, and the largest living gecko, *Rhacodactylus leachianus*, was eaten in some parts of New Caledonia. Not surprisingly, both of these examples come from islands, where there are no native mammals other than bats, and large geckos might be expected to provide as significant a source of protein as any forest animal.

Are parts of geckos used as medicine?

At least 14 species of geckos are used to treat various diseases and medical conditions in traditional medicine. Traditional medicine, based on the use of natural products, is widely practiced in developing countries where Western health care is unavailable or unaffordable, but it has also become popular as an alternative type of medical treatment in developed countries. Geckos are used in parts of Brazil, where they are cooked whole, powdered in tea, or used to make a broth to treat measles, sore throat, chickenpox, heart and liver ailments, and stroke and in the Kumaon Himalayas of India, where members of the Shoka tribes boil *Hemidactylus* in oil for the treatment of eczema. By far the most extensive use of geckos is in Chinese traditional medicine, where large geckos (*Gekko gecko* and *G. reevesii*) are used to treat a diversity of ailments including asthma, kidney stones, tuberculosis, and diabetes. Smaller species are employed in treating epilepsy, osteoporosis, and cancer. In recent decades, controlled clinical trials have been used to test the effectiveness of gecko-derived treatments and to identify the active ingredients in them, but the results have been inconclusive. Normally geckos are eviscerated and dried and then ground into a powder that can be mixed with other foods or taken as a tablet, either alone or with other ingredients. In Southeast Asia and in China, geckos are also used to make rice wine—sometimes with the gecko actually suspended in the bottle—that has supposed health benefits.

The Voracious Gecko (*Gehyra vorax*) of Fiji is one of relatively few geckos that was regularly eaten by people. Only relatively large geckos can provide even a small meal for a human.

Geckos: The Animal Answer Guide

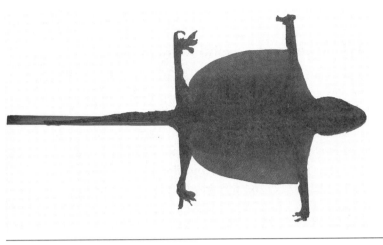

Geckos are widely used in traditional medicine to treat many ailments. Tokay Geckos (*Gekko gecko*) are especially valued in Chinese traditional medicine and are often sold eviscerated and dried. Demand for geckos in the medicinal trade has conservation implications.

The adhesive mechanism of geckos has inspired engineers and biologists to develop new technologies with applications in industry, medicine, and defense. One gecko-inspired material can support hundreds of pounds of suspended weights. From Bartlett et al. (2012). Courtesy of Michael Bartlett.

Can gecko-style adhesion be useful to humans?

The setae of gecko feet have inspired scientists and engineers to design artificial dry adhesives that function on the same principles. These take advantage of the fact that the frictional and van der Waals forces depend less on the type of material than on its physical attributes and mechanical performance. Thus different substances, like polymers and carbon nanotubules, have been used in gecko-inspired artificial adhesives. Some of the applications include bandages that might take the place of sutures in surgery and climbing robots that can conduct surveillance, clean or repair in hard-to-reach places, or explore dangerous or inhospitable areas, including those in space. Nanotechnologists are especially interested in replicating the "self-cleaning" property of gecko setae. The nature of the physical forces operating at the level of the setae results in any dirt and debris being left on the substrate rather than clogging the scansors. Incorporating this trait into any manmade gecko-inspired materials would help them maintain their adhesive properties through many uses, making them more cost-effective.

Chapter 9

Gecko Problems (from a human viewpoint)

Are geckos pests?

Whether geckos are pests or not depends on one's perspective. In areas of the world where geckos regularly come into houses, they are usually welcome. They eat insect pests, such as flies, mosquitoes, and moths as well as other invertebrates, like spiders, that many people do not like. However, the cost for these "gecko services" is that the homeowner may have to deal with gecko droppings. These may be deposited on walls or on furniture or inside drawers or cabinets. Although unsightly, gecko droppings, which normally include a fecal pellet that is blackish and an attached, white, pasty to solid urinary product, are small and relatively easy to cleanup.

Some diseases, most notably *Salmonella* (see below), can be transmitted by gecko feces, but in most cases the risks of disease from this source are less than from disease-carrying insects that the geckos eat. Geckos can also disrupt the quiet of a home with their vocalizations, but only a few geckos, such as the Tokay (*Gekko gecko*) are loud enough to be truly obtrusive. Most species of geckos that will live with humans are nocturnal and spend their days hidden behind pictures on the wall or inside cupboards. At night they dash out to capture prey or to interact with one another and are barely noticed by homeowners who are accustomed to them.

No gecko can be said to be a significant agricultural pest, but occasionally geckos are accidentally involved in food processing. For example, I once was asked to identify a gecko whose skull had been honey-roasted and included in an airline bag of almonds.

Do geckos have diseases and are they contagious?

Like all animals, geckos can suffer from a variety of ailments. Many of these are the result of poor housing conditions or inadequate nutrition and are not directly related to human health. However, geckos can carry certain zoonoses, infectious diseases that can be spread from animals to humans. The most well-known of these is the bacterium *Salmonella*, which is normally present in the digestive tract of many healthy lizards. It can be passed in the feces of geckos and can infect a person if ingested. This usually occurs if that people put their hands in their mouth or touch food following the handling of the animal itself or something else that has been in contact with the feces. *Salmonella* usually causes diarrhea, fever, and cramps in humans but can spread systemically and become life threatening.

Luckily, its spread to humans can easily be prevented by following basic sanitary rules, like hand washing. It is possible that some of the parasites carried by geckos, such as lung worms and certain ticks that can serve as disease vectors can survive in or on human hosts, but the likelihood of the accidental transfer from a gecko to a human is small. Cryptosporidiosis, a protozoan-caused disease that is often fatal in geckos, is caused by a species of *Cryptosporidium* different from any of those that are known to cause the disease of the same name in humans. Small, red trombiculid mites that are found on many geckos in the wild are also not dangerous for people. In

A "mite pocket" in the axilla (armpit) of a Kaokoveld Gecko (*Pachydactylus oreophilus*) contains a cluster of trombiculid mites. Such mites are common on wild geckos and usually cause no serious damage to their hosts, although heavy infestations can result in local skin irritation. Courtesy of Johan Marais.

Geckos: The Animal Answer Guide

some cultures geckos are believed to cause leprosy or other diseases affecting the skin, but such beliefs are unfounded.

How do I keep geckos away from my house?

Geckos in houses are largely self-limiting. Unless there is a large population of insects, only a few geckos are likely to be resident inside any particular building. However, if even this is too much, there are some steps that one can take to make a home or other building less attractive to geckos. Most importantly, a well-sealed building will generally exclude geckos, so screens on windows and doors are particularly effective. Avoiding open food and garbage containers will decrease both insect pests and their gecko predators. Some household pets, especially cats, are very good gecko catchers, although wary geckos that keep to ceilings and high walls may successfully avoid them. Keeping geckos away from outside building walls is more difficult, as they are attracted both by insects drawn to artificial lights and by convenient hiding and ovoposition sites offered by eaves, shutters, and other architectural features. Spraying pesticides for insects will either remove a gecko's prey base or kill the gecko through direct exposure to the poison or by the ingestion of poisoned insects. Pesticide use is not particularly eco-friendly but can be effective, although it requires frequent reapplication as surrounding populations of both insects and geckos will quickly recolonize, particularly in the tropics.

Chapter 10

Human Problems (from a gecko's viewpoint)

Are any geckos endangered?

Yes, a number of geckos are considered endangered. The IUCN (International Union for Conservation of Nature) maintains a Red List of threatened species, which in 2011 included 55 geckos, or less than 4 percent of recognized species, in the endangered or critically endangered categories (Table 10.1). The majority of these include species on islands like Madagascar (19 species) and New Caledonia (14 species), including many species of Day Geckos (*Phelsuma*) and Nimble Geckos (*Dierogekko*), respectively. However, this does not give an accurate view of all the geckos that are really endangered, as fewer than one-third of all geckos have been assessed by the IUCN and in some cases assessments are out of date. In addition, so little information is known about many geckos that their conservation status cannot be meaningfully assessed. Many countries have their own Red Lists, and these list many more species as endangered. As an example, the Salt Marsh Gecko (*Cryptactites peringueyi*), considered critically endangered in its South African homeland, has yet to be assessed by the IUCN. Other species may be doing well in parts of their range but not in others. For instance, Tokay Geckos have been considered endangered in China since 1998 but are not listed as threatened by other countries in which they occur.

What is the rarest gecko?

Many geckos are so poorly known that their rarity is hard to determine. Some species, such as the Indian Bent-toed Gecko, *Cyrtodactylus malcolm-*

The restricted ranges of many geckos place them at risk for extinction. The Néhoué Nimble Gecko (*Dierogekko nehoueensis*) lives in a small area of northern New Caledonia. All members of this genus are threatened or endangered, chiefly due to threats from habitat destruction through mining and other human uses.

smithi, are known only from one or two specimens collected more than a century ago. While scientists' failure to find these geckos again may mean that they are extinct, it more likely reflects our lack of knowledge about the distribution and biology of the species. This can be seen in the case of the New Caledonian Crested Gecko, *Correlophus ciliatus*. This distinctive species was first described in 1866 and was reported to be common for about a decade after its discovery. Then it was not seen again for more than 100 years, despite many searches, and was considered likely to be extinct. In 1992, it was rediscovered and since then has been found to be locally abundant. It has since become one of the most widely bred and kept of gecko species. Another New Caledonian giant gecko, *Rhacodactylus trachycephalus*, however, really does appear to be one of the rarest of all geckos. It has a very small distributional range, occurring only on one or two very small (less than 0.4 square miles or 1 square kilometer) offshore islands and has a low reproductive rate, giving live birth to at most one set of twins per year. These factors are coupled with major threats: the islands where they live have introduced populations of rats, which are known to prey on geckos, and the geckos are in high demand in the pet trade and are easily accessed by smugglers.

Have any geckos become extinct because of humans?

The giant *Hoplodactylus delcourti* of New Zealand is probably extinct. Its demise was probably the result of a combination of natural causes (the decrease in the range of giant kauri trees due to climate change over the

Table 10.1. IUCN (International Union for the Conservation of Nature) list of geckos in the three threatened status categories. Two other geckos (*Hoplodactylus delcourti* and *Phelsuma gigas*) are listed as extinct by the IUCN. Species occurring on islands are indicated by an asterisk (*). Recent changes in taxonomy have resulted in the movement of *Rhacodactylus ciliatus* and *R. sarasinorum* to the genus *Correlophus*, *R. chahoua* to *Mniarogekko*, and *Cyrtopodion amictophole* to *Mediodactylus*.

Family	VU (vulnerable)	EN (endangered)	CR (critically endangered)
Carphodactylidae		*Nephrurus deleani*	*Phyllurus gulbaru*
Diplodactylidae	*Correlophus ciliatus**	*Bavayia exsuccida**	*Dierogekko inexpectatus**
	*Mniarogekko chahoua**	*Bavayia goroensis**	*Dierogekko kaalaensis**
	*Toropuku stephensi**	*Bavayia ornata**	*Dierogekko koniambo**
		*Correlophus sarasinorum**	*Dierogekko nehoueensis**
		*Dierogekko validiclavis**	*Dierogekko poumensis**
		*Eurydactylodes symmetricus**	*Dierogekko thomaswhitei**
		*Rhacodactylus trachyrhynchus**	*Eurydactylodes occidentalis**
			*Oedodera marmorata**
Pygopodidae	*Aprasia rostrata*		*Aprasia aurita*
	Delma impar		
	Delma labialis		
	Delma torquata		
	Ophidiocephalus taeniatus		
	Paradelma orientalis		
Eublepharidae		*Goniurosaurus kuroiwae**	
Gekkonidae	*Ailuronyx trachygaster**	*Ebenavia maintimainty**	*Cnemaspis anaikatatiensis*
	*Blaesodactylus boivini**	*Gehyra barea*	*Hemidactylus dracaenacolus**
	*Christinus guentheri**	*Luperosaurus joloensis**	*Lygodactylus mirabilis**
	*Cyrtodactylus cavernicolus**	*Luperosaurus macgregori**	*Matoatoa spannringi**
	*Gekko ernstkelleri**	*Lygodactylus intermedius**	*Paroeduralo hatsara**
	*Gekko gigante**	*Lygodactylus ornatus**	*Phelsuma antonosy**
	Gekko swinhonis	*Lygodactylus roavolana**	*Phelsuma masohoala**
	*Lepidodactylus listeri**		*Phelsuma pronki**
	*Lygodactylus bivittis**	*Mediodactylus amictopholis*	
	*Lygodactylus blanci**	*Paragehyra gabriellae**	
	*Lygodactylus gravis**	*Paroedura masobe**	
	*Lygodactylus madagascariensis**	*Paroedura sanctijohannis**	
	Lygodactylus methueni	*Paroedura tanjaka**	
	*Matoatoa brevipes**	*Phelsuma flavigularis**	
	*Nactus coindemirensis**	*Phelsuma guentheri**	
	*Nactus serpensinsula**	*Phelsuma klemmeri**	
	*Paragehyra petiti**	*Phelsuma robertmertensi**	
	*Paroedura androyensis**	*Phelsuma roesleri**	
	*Paroedura vazimba**	*Phelsuma seippi**	
	*Phelsuma breviceps**	*Phelsuma serraticauda**	
	*Phelsuma hielscheri**	*Phelsuma vanheygeni**	
	*Phelsuma nigristriata**	*Uroplatus guentheri**	
	*Phelsuma standingi**	*Uroplatus malahelo**	
		*Uroplatus pietschmanni**	

Family	VU (vulnerable)	EN (endangered)	CR (critically endangered)
Gekkonidae	*Pseudogekko brevipes**		
	Uroplatus ebenaui*		
	Uroplatus giganteus*		
	Uroplatus henkeli*		
	Uroplatus malama*		
Phyllodactylidae	*Phyllodactylus leei**		
Sphaerodactylidae	*Saurodactylus fasciatus**	*Sphaerodactylus armasi**	*Gonatodes daudini**
	Sphaerodactylus callocricus*	*Sphaerodactylus micropithecus**	*Sphaerodactylus williamsi**
	Sphaerodactylus kirbyi*	*Sphaerodactylus pimienta**	
	Sphaerodactylus scapularis	*Sphaerodactylus storeyae**	
	Sphaerodactylus torrei*		

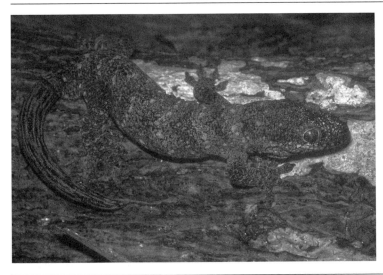

The Lesser Rough-snouted Giant Gecko (*Rhacodactylus trachycephalus*) from New Caledonia may be the rarest gecko in the world. It has a small insular distribution, produces at most two offspring per year, and is sought after for the pet trade.

Courtesy of Mark O'Shea.

past 3,000 years) and human ones (land use changes and logging of kauri following the European colonization of New Zealand). A clearer example is that of *Phelsuma edwardnewtoni*, a large Day Gecko from Rodrigues in the Mascarene Islands in the Indian Ocean. This gecko is known from only six specimens, all collected before 1920. The ecology of its small (43 square miles or 109 square kilometers) island home was dramatically changed when humans introduced rats and cats, which were responsible for the gecko's demise. A similar fate probably befell *P. gigas*, a huge (7.5 inch/190 millimeter head and body length) nocturnal gecko, also from Rodrigues. It had become rare by the 1760s and has not been seen since 1841. Numerous expeditions to Rodrigues and its even smaller satellite islands confirm that both species are extinct. The Mascarenes have seen perhaps more recent vertebrate extinctions than any place else on earth, having lost giant tortoises, a boa, and many birds, including the dodo and the solitaire, to human activity. It is probable that other geckos, mostly on islands, have gone extinct as a result of the introduction of invasive predators or habitat destruction before they were even described.

Are geckos protected by law?

Whether geckos are protected depends both on their conservation status and where they live. In most countries native reptiles, including geckos, have some form of protection and special permission is required to collect or keep them. These regulations vary significantly from country to country, even from state to state or province to province. In most cases, geckos that are considered to be endangered receive the greatest protection. Collecting, killing, or harming protected geckos can result in fines or even imprisonment. Some geckos are protected under CITES (Convention on International Trade in Endangered Species of Wild Fauna and Flora), an agreement between governments that protects endangered species by controlling their international trade. No geckos are listed in the most endangered category of CITES, but all species of *Phelsuma* and *Uroplatus*, as well as *Nactus serpensinsula*, the Serpent Island Gecko from

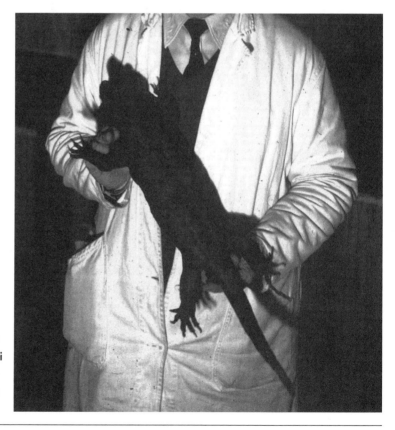

The largest gecko that ever lived, known only from one stuffed specimen, was Delcourt's Giant Gecko (*Hoplodactylus delcourti*) from New Zealand. Human destruction of kauri tree forests, along with the introduction of rats, may have contributed to its extinction.

Geckos: The Animal Answer Guide

the Mascarene Islands, are listed in CITES Appendix II (species that may become threatened by extinction unless trade is closely controlled). Even common *Phelsuma* species are included in this category, because they are similar enough in appearance to their endangered relatives that they could be confused for one another. All of the geckos of New Zealand are protected under another category of CITES.

Will geckos be affected by global warming?

Global warming is a reality. There is ample evidence to suggest that most organisms, particularly ectotherms, may be affected by it in the not-too-distant future, although exactly how remains uncertain. A rise in environmental temperature could have short-term benefits for at least some nocturnal geckos. Most night-active geckos that have been studied are known to be active at temperatures below their preferred and presumably physiologically optimal temperatures. The difference is due to the inability of geckos to maintain their higher preferred body temperatures as ambient temperatures fall after dark. In a world with warmer temperatures, some geckos could benefit, as long as they could avoid overheating during the day. Diurnal geckos, however, might be expected to suffer in the same way predicted for other lizards. Several different scenarios have been proposed. In one, higher temperatures would force geckos into retreat sites for longer periods, decreasing available foraging time and thus decreasing the energy stored. In the worst case geckos would not have sufficient energy to reproduce and could become locally extinct. Geckos at higher latitudes might be at higher risk because these areas are predicted to suffer the greatest increase in temperature over time. Likewise, montane geckos, already living in the coolest environments available to them would have nowhere cooler to move if local temperatures increased. However, geckos living in tropical lowlands, where temperatures are high and relatively stable across seasons and times of day might also be at high risk. Although their ranges might move polewards or up in elevation over time, these lizards tend to have a narrower range of acceptable temperatures than more cool-adapted species and are already living closer to their thermal optimum. Under a scenario of rapid warming, even of only a few degrees, these geckos might no longer be able to survive.

What are the greatest threats to geckos?

Geckos are at risk from the same sources that affect most other wildlife. Probably the single greatest risk to geckos is habitat destruction. Most geckos have highly restricted areas of occurrence, often only a sin-

Global warming may temporarily benefit some nocturnal geckos that must now function under cooler-than-preferred temperatures, but diurnal species and cool-adapted species from high elevations or high latitudes, like the New Zealand Striped Gecko (*Toropuku stephensi*), could become extinct. © Tony Whitaker.

gle mountaintop or small island. Within this small area, there may still be many suitable retreat sites for geckos and some human activity, even if destructive, may be tolerated, but not if the entire area is under wholesale threat from mining or logging. The New Caledonian Nimble Geckos of the genus *Dierogekko* are a good case in point. Study of this genus has revealed that nearly every mountaintop in northwestern New Caledonia supports a unique species. Most of these mountains have been mined, are being mined, or are under license to be mined for nickel or other metals. The mining removes virtually all of the vegetative cover and thus removes the entire habitat for these species.

Another major threat is the international trade in geckos. Every year millions of geckos are imported into the United States alone. Geckos are attractive animals and in recent decades have become among the most popular of all exotic pets. Most of the commonly kept species, even if they were once illegally collected from the wild, are now bred easily and cheaply in captivity and pose no major threat to wild populations. However, some of the rarest species of geckos are also among the most desirable. There is a demand for illegally collected or exported species from the ever-growing exotic pet trade. More significant is the trafficking in geckos for traditional medicinal uses. Tokay Geckos (*Gekko gecko*) are in such demand to meet the requirements of the Chinese traditional medicine market that they have become endangered in China itself and are now imported from throughout the rest of Asia, where overcollecting has caused localized habitat destruction and depleted populations.

Geckos: The Animal Answer Guide

Chapter 11

Geckos in Stories and Literature

What roles do geckos play in religion and mythology?

It is difficult to determine the role of the gecko in religion, as many societies did not or do not distinguish between types of lizards or recognize geckos as a distinctive type of lizard. One might expect that the Bible, having originated in the Middle East—an area of at least moderate gecko diversity—might say something about them. But reference to the most famous English version of the Bible, the Cambridge Edition of the King James Bible (1611), reveals no mention of geckos. However, other reptiles are noted in that part of the Bible in which the Lord tells Moses and Aaron about clean and unclean foods:

> These also shall be unclean unto you among the creeping things that creep upon the earth; the weasel, and the mouse, and the tortoise after his kind, and the ferret, and the chameleon, and the lizard, and the snail, and the mole. These are unclean to you among all that creep: whosoever doth touch them, when they be dead, shall be unclean until the even.
>
> Leviticus 11:29–31 (King James Bible)

Of course, when the Bible reached Europe in the Christian Era, some of the animals mentioned were unknown. The original Hebrew (and later Greek and Latin) Bible texts were translated in such a way as to make sense to the people who would read them in their own languages. In the King James Bible, the word "ferret" was provided as the translation for a word referring to an animal that calls out or cries, and which, in modern

115

Hebrew, is used for "gecko." Thus, a more accurate translation of this part of Leviticus reads:

> Of the animals that move about on the ground, these are unclean for you: the weasel, the rat, any kind of great lizard, the gecko, the monitor lizard, the wall lizard, the skink and the chameleon. Of all those that move along the ground, these are unclean for you. Whoever touches them when they are dead will be unclean till evening.
>
> Leviticus 11:29–31 (New International Version, 1984)

So a gecko does appear once in the Bible, but in a rather negative context. Unfortunately, geckos fare no better in Islamic tradition, which otherwise instructs Muslims to treat animals with compassion. The Book on Salutations and Greetings (Kitab As-Salam) in the Sahih Muslim (a major collection oral traditions relating to the words and deeds of the Prophet-Muhammad in Sunni Islam) includes a chapter entitled "The Desirability of Killing a Gecko." This states that Muhammad instructed people to kill geckos, which he considered noxious creatures: "He who killed a gecko with the first stroke for him are ordained one hundred virtues, and with the second one less than that and with the third one less than that." Geckos were supposedly singled out for their treachery. During the Hijra (Muhammad's journey from Mecca to Medina in AD 622 and the first year of the Islamic calendar), Muhammad spent 3 days hiding from his persecutors in the Cave of Thur (Ghar al-Thawr). According to tradition, a gecko tried to give away his position by calling out to his enemies.

Although there is no mention, of which I am aware, of geckos in connection with Zoroastrianism, a religion and philosophy that originated in Persia in the sixth century BC, there is nonetheless a connection between the two. In the original description of the large eublepharid gecko *Eublepharis angramainyu* from Iran and Iraq, Steve Anderson and Alan Leviton stated that the species name is "derived from 'Angra Mainyu' the Zoroastrian 'Spirit of Darkness,' in reference to the nocturnal habits of these animals." Angra Mainyu has also been interpreted as an evil spirit or a deceiving false god and has been equated to Satan in the Christian tradition.

In the Aboriginal cultures of Australia, geckos play a more positive and substantial role. Ipilya is the spirit of a gigantic gecko who, in the mythology of the people of Groote Eylandt, Northern Territory, brings on the monsoon through his actions. In another story of the Dreamtime, a gecko-man is credited with helping to create the day and night. In honor of this, geckos were not to be harmed. Chambers Pillar, a sandstone formation in central Australia is said to be the gecko ancestor Itirkawara, who was turned to stone for his disgraceful behavior. Geckos are represented in rock art and petroglyphs in Australia and in other parts of the world, including Sri Lanka, where the Vedda people drew images of the Golden Gecko, *Calodactylodes illingworthorum*.

To the Maori of New Zealand, geckos and other reptiles were the descendants of Punga, himself a son of the sea god Tangaroa. Punga's offspring were all considered to be ugly. They were also associated with Whiro, the god of darkness, evil, and death. In this context geckos were feared, and the vocalizations of green geckos were interpreted as evil omens. However, geckos could also serve as guardians to watch over the dead or to protect a building or important possessions. Similar beliefs about geckos were held in most of Polynesia, including Hawaii, where geckos were regarded as representatives or manifestations of *Mo'o*, a powerful dragon-like guardian spirit. Because of this association geckos are regarded as good luck and to kill one is bad luck. Geckos in Polynesia are also often associated with a link between the living and the dead or between generations. In New Caledonia, part of Melanesia, geckos are both feared and respected. They serve as totems for particular tribes and can be associated with spirits of people's ancestors.

Geckos play little or no role in the major modern religions but have figured prominently in several non-Western belief systems and mythologies. Geckos are depicted in the rock art of several cultures, including that of the Vedda people of Sri Lanka.

Courtesy of Anslem de Silva.

What roles do geckos play in native cultures?

In many cultures geckos are associated with luck, either good or bad. In Sri Lanka there is a long tradition of making predictions based on the actions of geckos. Even today an annual almanac is published that contains the sections "predictions based on gecko calls" and "predictions based on the body area on which a gecko falls." In much of East and Southeast Asia, geckos are typically associated with good luck and increased fertility. Probably because of these associations, geckos are especially important ingredients in traditional medicine as practiced in China, Vietnam, and other countries.

However, in parts of the Arab world and South Asia, they have been blamed for the spread of disease, specifically leprosy, leukoderma (vitiligo), and other skin conditions. This almost certainly stems from the observation of geckos in the process of shedding their own skins. In parts of Pakistan and Afghanistan, the tail, skin, and blood of geckos are used in the preparation of potions to weaken one's enemies, and in Zimbabwe geckos are ingredients in "husband taming" preparations. It is striking to note that the belief that geckos are venomous or poisonous seems to transcend all boundaries. In Amazonia, the Barasana people consider the Turnip-tailed Gecko, *Thecadactylus rapicauda*, to be venomous. They also believe that poison is injected through the gecko's feet. The Mayans believe that geckos can envenomate by dropping their tails and that the eublepharid, *Coleonyx*, can kill a human merely by touch. And in parts of South Africa, a gecko bite is believed to bring on hysteria, resulting in uncontrollable laughter and eventually death. Such beliefs mirror those presented in the Bible, which states that any food or water touched by a gecko is unclean.

In the Maori culture of New Zealand, geckos, and lizards in general, play an important role in belief systems and in art. They were sometimes carved on the ridgepoles or gables of buildings, probably in association with their role as guardians. They also appeared on burial chests, containers of human bones, and on tombs, also serving as guardians and as reminders of their link to Whiro, the god of darkness. Geckos also appear in Maori rock art and on greenstone and bone carvings. They are also prominent in the visual arts of other Polynesian peoples. Given their roles in religion and mythology, geckos also appear in the carvings of New Caledonia and the Aboriginal designs of Australia.

As in Oceania, geckos are also perceived in a generally positive light in the Philippines and on Bali (a chiefly Hindu island in predominantly Islamic Indonesia). In both of these cultures, geckos play a role in traditional stories and fables. Some of these stories have been translated and published in the West under such titles as *Go to Sleep Gecko: A Balinese Folktale*, *Gecko's Complaint: A Balinese Folktale*, and *Tuko and the Birds: A Tale from the Philippines*.

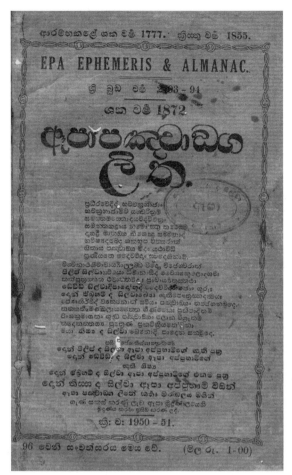

Geckos are associated with luck, either good or bad, in many cultures. In Sri Lanka, annual almanacs list "predictions based on gecko calls" and "predictions based on the body area on which a gecko falls." Courtesy of Anslem de Silva.

What roles do geckos play in popular culture?

Beginning in the late twentieth century, geckos began to creep (no pun intended) into Western popular culture. Geckos were first used in advertising in places like Australia, Bali, the Philippines, and Hawaii, where the lizards themselves occur and are known to locals. By the end of the century, geckos were everywhere. Stylized geckos initially appeared on tee-shirts, sportswear, surfing products, and climbing gear. Today geckos are in the names and logos of thousands of companies selling everything from furniture to computer accessories. There are Gecko Rental Car companies (one each in Mexico and Namibia); countless restaurants, pubs, and bars with "gecko" in their names; and brands of gecko vodka, alcohol-free liquors, and wines. Sports teams using the gecko name include the Bali Gecko Australian Rules Football Club, the George Washington High School Geckos (Guam), the GateWay Community College Geckos, and the Shadow Mountain Geckos Boys 11 and under Baseball Team (both in Phoenix, Arizona), the Laie (Hawaii) Elementary School Geckos, and the

Geckos are widely used as symbols in marketing, music, and sports. In areas of the tropics where geckos are conspicuous, gecko-themed businesses are common, like this hotel and restaurant in Sri Lanka. Courtesy Anslem de Silva.

Albuquerque (and then Sacramento) Geckos professional soccer team (now defunct). And although the Seasiders is the name of the Brigham Young University–Hawaii sports teams, their mascot is Kimo the Gecko.

In the world of music, artists named after geckos represent almost all genres of music. They include the following: "Gecko," "The Gecko Brothers," "Unified Gecko," "Gecko's Tear," "Gecko Turner," "Gecko Temple," "Gecko Yamori," "DJ Gecko," "GekkoProjekt," "Seth Gekko," "Pinkeye d'Gekko," and "DJ Gordon Gekko." The number of songs, CDs, video games, websites, and blogs using some variant of "gecko" in their titles is staggering.

One of the most well-known geckos in popular culture is Gordon Gekko, a character played by the actor Michael Douglas in the hit 1987 movie *Wall Street* (and again in 2010's *Wall Street: Money Never Sleeps*). Gekko's famous phrase was "greed, for the lack of a better word, is good." The phrase and the character have become emblematic of the excesses of the 1980s and receive many secondary pop culture references. In 1993 herpetologist Indraneil Das described *Cnemaspis* (now transferred to *Cyrtodactylus*) *gordongekkoi* from the island of Lombok, Indonesia, as an homage to the character.

Certainly the most ubiquitous pop culture gecko in the United States of the late twentieth and early twenty-first century is the GEICO Gecko, a spokesman (and registered trademark) for the auto insurance company GEICO in many of its television commercials. It first appeared in 1999 and has remained a mainstay of GEICO's popular and iconic advertising campaigns. Although the gecko is highly anthropomorphized and anatomically incorrect (for example, he has movable eyelids instead of a brille and his first digit is like a human thumb rather than being greatly reduced), he

Geckos: The Animal Answer Guide

is clearly based on a Day Gecko (*Phelsuma*). GEICO insures only in the United States, but somehow the GEICO Gecko has a Cockney accent (often mistaken for Australian by Americans).

What roles have geckos played in poetry and other literature?

The relatively recent awareness of geckos in Western culture limits the role that they play in literature. There are no classic novels that feature geckos, nor are they used allegorically in great poetry. Nonetheless, geckos do feature in some literary works. They have been the subject of several poems, including "The Gecko" by the Australian poet Leon Gellert (1892–1977), a portion of which serves as the epigraph for this book. An examination of the Internet reveals many amateur poems involving geckos as well. Many of these, like Gellert's poem, remark on the revulsion felt by the observer at seeing a gecko, although in most poems the gecko is ultimately regarded favorably, usually because it eats insects. Geckos also have found a place in children's literature. Bruce Hale has authored *How the Gecko Lost His Tail*, *The Story of the Laughing Gecko: A Hawaiian Fantasy*, *Moki and the Magic Surfboard: A Hawaiian Fantasy*, and a series of books featuring Chet Gecko, a gecko who is also an elementary school detective. Gill McBarnet's *The Goodnight Gecko* is also a popular children's story.

Chapter 12

"Geckology"

Who studies geckos?

Although people are interested in geckos and study captives in their own homes for personal enjoyment, there are not many professional gecko biologists. This is partly because some parts of the world that have the most geckos tend not to have many researchers and partly because geckos, often being small and nocturnal, can be quite difficult to study. Much of the scientific work done on geckos is in the field of systematics, which seeks to document biodiversity and is carried out by taxonomists, who discover, describe, and classify geckos. Systematists also use data from morphology and from DNA sequences to construct phylogenetic trees, or hypotheses of relationship among geckos. Other scientists, including ethologists and ecologists, study the interactions of geckos with one another and with other organisms in their environments, whereas morphologists and physiologists investigate the form and function of geckos. Still others focus on sensory biology or reproductive biology. Most published gecko research is carried out at universities and museums, but zoo-based biologists and serious herpetoculturalists have been especially important in identifying aspects of behavior and reproduction, while exotic pet and reptile veterinarians have learned much about gecko health, diseases, and treatments.

Many herpetologists have contributed greatly to our knowledge of geckos. Carolus Linnaeus initiated the system of scientific names now in use and so coined the first such names for geckos. However, Linnaeus was not fond of amphibians and reptiles, and it was left to later workers to distinguish geckos from other lizards and from salamanders, with which they had often been confused. Later in the nineteenth century, many museum-based systematists described geckos from all over the world. These included herpetologists from the largest European museums, like those in London, Paris, and Berlin. By the early twentieth century both gecko taxonomy and anatomy were well studied (Table 12.1), but gecko ecology, behavior, genetics, and other fields remained largely unstudied until after World War II. Today there are many gecko specialists who are actively publishing new findings in all disciplines of biology all the time. A glance through the references section of this book will provide the names of some, but by no means all, of the most important such specialists. Herpetocultural contributions to the study of geckos in captivity have also mostly been recent. Indeed, in the last several decades, the keeping of geckos has gone from "Dark Ages" to high tech.

Table 12.1. Selected biologists who made important contributions to the study of geckos from the mid-eighteenth century to the mid-twentieth century. Most were taxonomists and/or anatomists. Other areas of gecko study began in earnest only about 1950 (see References for citations to a diversity of topics of study by modern gecko researchers).

Name	Country	Life dates	Contributions
Carolus Linnaeus	Sweden	1707–1778	First gecko descriptions using two part Latin name
Martinus Houttuyn	Holland	1720–1798	Early contributors to the recognition of
Johann Gottlob Schneider	Germany	1750–1822	geckos as a distinct group of lizards
Wilhelm Gottlieb Tilesius von Tilenau	Germany/Russia	1769–1857	
André-Marie-Constant Duméril	France	1774–1860	
John Edward Gray	United Kingdom	1800–1875	Nineteenth-century museum scientists
Gabriel Bibron	France	1806–1848	who described many geckos from
Auguste-Henri-André Duméril	France	1812–1870	European colonies around the world
Wilhelm K. H. Peters	Germany	1815–1883	
Albert C. L. G. Günther	Germany/United Kingdom	1830–1914	
George Albert Boulenger	Belgium/United Kingdom	1858–1937	
Robert Wiedersheim	Germany	1848–1923	Contributors to the study of gecko
Friedrich Siebenrock	Austria	1853–1925	anatomy
Margarete Vera Wellborn	Germany	1909–????	
Franz J. M. Werner	Austria	1867–1939	Systematic herpetologists who
Malcolm A. Smith	United Kingdom	1875–1958	described new geckos and studied
Thomas Barbour	USA	1884–1946	their evolution
Arthur Loveridge	United Kingdom/USA	1891–1980	
Garth Underwood	United Kingdom	1919–2002	

Which species are best known?

It is probably true to say that almost no gecko species are well known in the wild, but several species are quite well known in the captive or laboratory setting. Appropriately, the Tokay Gecko is perhaps the most well studied of all. Because it is the species that gave its name to the entire family of Gekkonidae, because it has a broad distribution from northeast India and Bangladesh to Indonesia, and because it is a very large species (maximum of 7 inches/180 millimeters in head and body length) that can be easily observed and instrumented, the tokay has been the focus of hundreds of investigations. These have covered the fields of genetics, karyology, behavior, ecology, vocalization and audition, physiology, reproduction, vision, hearing, locomotion, feeding, endocrinology, and others. It has only recently been demonstrated, however, that several species have been masquerading under the name *Gekko gecko* and that some of what we thought we knew about this species actually applies to *Gekko reevesii*, the so-called Black Tokay.

Another species that is well known is the Leopard Gecko (*Eublepharis macularius*). This species has been bred for more than 30 generations in captivity. It is estimated that the number of hatchlings produced per year in the United States alone may be in the hundreds of thousands. Indeed, this is one of the most popular "starter pets" in the United States and Europe. Much is known about its care and maintenance, and its reproductive biology is well-studied. However, its biology in its native range (eastern Iran to northwestern India) is very poorly known.

Second only to the Leopard Gecko in terms of knowledge based on captive animals is the Crested Gecko, *Correlophus ciliatus*. All of the Giant New Caledonian Geckos are sought after as pets, but this species has a winning combination: it is spectacular in appearance, inexpensive to buy, and easy to keep and breed. As recently as the early 1990s, it would have been considered one of the least well-known geckos. After its rediscovery in New Caledonia and intensive captive breeding in Europe and America, it became familiar to every herpetoculturalist. Like the Leopard Gecko, it has been the subject of several books and can be purchased in any of several beautiful "morphs"—the result of selective breeding for attractive traits. Although now known to be locally abundant in the wild, its natural history remains poorly studied.

Hundreds of other species of geckos are kept in captivity and are known with respect to the basics of their biology. Geckos in the genera *Uroplatus* and *Phelsuma* are especially popular. Although members of both genera have been studied in the wild, for most species diet, life history traits, and physiology in the wild remain unknown. The best-studied geckos in the wild are those species that have been the focus of dedicated research, often based on years of fieldwork. These include some of the easier to observe "house geckos," like *Hemidactylus frenatus*, *H. turcicus*, *H. mabouia*, *H. platyurus*, and *Lepidodactylus lugubris*, as well as the three Israeli species of *Ptyodactylus*, the terrestrial African Barking Gecko *Ptenopus garrulus*, and the Western Banded Gecko of the American southwest, *Coleonyx variegatus*. Thanks to the work of H. Robert Bustard, Eric Pianka, Hal Cogger, Klaus Henle and others, Australian geckos, as a group, may be better known than those of any other region. Some of the best known of these are *Christinus marmoratus*, *Gehyra variegata*, *Heteronotia binoei*, and *Strophurus williamsi*. Laurie Vitt and colleagues, working in the Neotropics, have also revealed much about geckos in the genera *Thecadactylus*, *Gonatodes*, *Coleodactylus*, and *Gymnodactylus*.

Which species are least known?

One is tempted to say "most of them" because field-based data are available for a small minority of geckos. The biology of probably fewer than 5 percent could be said to be well characterized; that is, no aspect of their biology has been investigated. Indeed, for most geckos the only published literature is the species' original description and perhaps its inclusion in regional checklists or a phylogenetic tree. Ecological data are typically limited to the characterization of the habitat where the animal has been collected or perhaps an anecdotal record of diet or reproduction. Still, some geckos are more poorly known than others. Some of the gecko genera that are probably most unfamiliar, even to professional herpetologists, include *Perochirus* from Micronesia, *Haemodracon* from Socotra Island, *Pseudogekko* from the Philippines, *Cryptactites* from South Africa, and the pygopodid *Ophidiocephalus* from Australia. However, geckos from parts of the world that have been herpetologically understudied surely lead the list of least known. Among the understudied areas are parts of Equatorial Africa, Somalia and surrounding areas, and many regions of Asia, from northern Pakistan across India to Southeast Asia and Indonesia. Thankfully, in some of these areas there is ongoing field research and local scientists are collecting new data about the biology of described geckos and discovering new species.

Certainly, among the most poorly known geckos must be the many *Hemidactylus* species that are endemic to the Horn of Africa, the part of the continent that projects eastward toward the Arabian Peninsula. A number of these geckos were described during the Italian colonial period in Somalia and have not been reported on since—a situation exacerbated by the political instability of the region. For some of these species, all we have is a type specimen and a short description. But this situation is not unique. *Cyrtodactylus malcolmsmithi* was described from two specimens that were

Many geckos are restricted to small and often remote areas and are little studied. The Rough-snouted Socotran Leaf-toed Gecko (*Haemodracon trachyrhinus*) occurs only on the island of Socotra off the Horn of Africa and very little is known of its biology. Courtesy of Jon Boone.

The tiny Salt Marsh Gecko (*Cryptactites peringueyi*) of South Africa was only known from two type specimens for about 100 years, despite living near the large city of Port Elizabeth. When rediscovered in the 1990s, it was found to live in and around estuaries, a rather un-gecko-like habitat. It is one of hundreds of geckos that we know very little about. Courtesy of Bill Branch.

collected in northwestern India in the 1870s. Despite there being a specific locality for them, no new specimens have ever been found. No biological information is available about the species because it was only described in 1947, long after the death of the only people who ever saw it alive. Even more extreme is the case of *Cnemaspis boiei*. This species was described in 1842, from India (no more specific details), in just five lines of text. It, too, has never been found again.

How do scientists tell geckos apart?

Many different morphological structures vary among geckos and can be used to distinguish species from one another. The presence of movable

Geckos: The Animal Answer Guide

eyelids separates all eublepharids from other geckos and, of course, the absence of limbs identifies pygopodids. The structure of the toes provides a great deal of information and is often used to identify the genus to which a gecko belongs. This is reflected in many gecko names like *Gymnodactylus* (naked-toed), *Cyrtodactylus* (bent-toed), *Pachydactylus* (thick-toed), *Rhacodactylus* (ragged-toed), *Phyllodactylus* (leaf-toed), and *Ptyodactylus* (fan-toed). The reduction of the first digit of the feet and the presence of webbing between the toes are other toe-related clues to a gecko's identity. For monotypic gecko genera or for genera with only a few species, identification is usually quite straightforward. However additional information is needed to identify species within the larger genera. Among the types of characteristics that are often useful are the presence, shape, and number of rows of enlarged dorsal tubercles; the presence of ventrolateral folds along the flanks and whether these have denticles or spines on them; the type of scales underneath the original tail (broadly expanded, somewhat enlarged, or similar to the dorsal scales of the tail); the presence, number, and configuration of precloacal or femoral pores in adult males; and the configuration of the head scales, particularly those around the nostril, on the snout, under the chin, and along the lips. In many groups, color pattern is also an important feature in distinguishing geckos from one another. Usually adult color patterns, like bands or stripes or blotches, are useful in this regard. However, in some *Pachydactylus* species, adults all have similar patterns and hatchling coloration provides information that determines species. The genus *Cyrtodactylus*, for example, has more than 170 recognized species, so all of these features must be used in combination to unambiguously separate the taxa from one another.

In some cases, genetic data reveal the existence of previously unrecognized diversity within a gecko genus. For example, very similar looking geckos may have a high level of divergence in their DNA sequences. This can be an indication that cryptic species are present. Cryptic species are those that are biologically distinct from one another but may be nearly identical in appearance. In reality cryptic species can often be separated on morphological grounds, but the features that distinguish them may be so subtle that they go unnoticed until molecular data reveal that more than one species is present. A good example comes from the New Caledonian Nimble Geckos, *Dierogekko*. All members of the genus look very much alike, but when DNA sequence data identified eight different lineages, it was possible to go back to the specimens and distinguish them on the basis of small differences in size, precloacal pore number and organization, and color.

Differences in activity period, reproductive timing, vocalizations, and habitat preferences are all also useful adjuncts in telling geckos apart, especially when highly similar species occur in the same region.

Appendix A

Geckos of the World

This list includes extant (living) and recently extinct species of geckos recognized as valid as of 6 November 2012. † = recently extinct species, (VU) = Vulnerable, (EN) = Endangered, (CR) = Critically Endangered; * = CITES listed species. *Note:* The species composition of most gecko genera is very stable, but the membership of several Palearctic gecko genera (*Cyrtopodion, Mediodactylus, Altiphylax, Tenuidactylus*) remains problematic and is likely to change with more study.

Family Carphodactylidae
Carphodactylus: Chameleon Gecko; 1 species; northeastern Australia
   C. laevis
Nephrurus: Knob-tailed Geckos; 9 species; northeastern Australia
   N. amyae, N. asper, N. deleani (EN), *N. laevissimus, N. levis, N. sheai, N. stellatus, N. vertebralis, N. wheeleri*
Orraya: Long-necked Leaf-tailed Gecko; 1 species; Australia
   O. occultus
Phyllurus: Broad-tailed Geckos; 9 species; Australia
   P. amnicola, P. caudiannulatus, P. championae, P. gulbaru (CR), *P. isis, P. kabikabi, P. nepthys, P. ossa, P. platurus*
Saltuarius: Leaf-tailed Geckos; 6 species; north and central eastern Australia
   S. cornutus, S. kateae, S. moritzi, S. salebrosus, S. swaini, S. wyberba
Underwoodisaurus: Thick-tailed Geckos; 2 species; southern Australia
   U. milii, U. seorsus
Uvidicolus: Border Thick-tailed Gecko; 1 species; central eastern Australia
   U. sphyrurus

Family Diplodactylidae
Amalosia: Dwarf Velvet Geckos; 4 species; northern and northeastern Australia
   A. jacovae, A. lesuerii, A. obscura, A. rhombife
   Bavayia: New Caledonian Forest Geckos; 12 species; New Caledonia, Loyalty Islands (South Pacific)
   B. crassicollis, B. cyclura, B. exsuccida (EN), *B. geitaina, B. goroensis* (EN), *B. montana, B. nubila, B. ornata* (EN), *B. pulchella, B. robusta, B. sauvagii, B. septuiclavis*
Correlophus: New Caledonian Crested Geckos; 3 species; New Caledonia
   R. belepensis, R. ciliatus (VU), *R. sarasinorum* (VU)
Crenadactylus: Clawless Gecko; 1 species; central and western Australia
   C. ocellatus
Dactylocnemis: Pacific Gecko; 1 species; northern New Zealand and islands offshore
   *D. pacificus**

Dierogekko: New Caledonian Nimble Geckos; 8 species; New Caledonia
  D. inexpectatus (CR), *D. insularis, D. kaalaensis* (CR), *D. koniambo* (CR),
  D. nehoueensis (CR), *D. poumensis* (CR), *D. thomaswhitei* (CR), *D. validiclavis* (EN)
Diplodactylus: Stone Geckos; 18 species; southwestern Australia
  D. calcicolus, D. capensis, D. conspicillatus, D. fulleri, D. furcosus, D. galaxias,
  D. galeatus, D. granariensis, D. kenneallyi, D. klugei, D. mitchelli, D. ornatus,
  D. polyophthalmus, D. pulcher, D. savagei, D. tessellatus, D. vittatus, D. wiru
Eurydactylodes: New Caledonian Chameleon Geckos; 4 species; New Caledonia
  E. agricolae, E. occidentalis (CR), *E. symmetricus* (EN), *E. vieillardi*
Hesperoedura: Southwestern Velvet Gecko; 1 species; Australia
  H. reticulata
Hoplodactylus: New Zealand Giant Geckos; 2 species; New Zealand offshore islands
  H. delcourti†, H. duvaucelii*
Lucasium: Ground Geckos; 11 species; Australia
  L. alboguttatum, L. bungabinna, L. byrnei, L. damaeum, L. immaculatum, L. maini,
  L. occultum, L. squarrosum, L. steindachneri, L. stenodactylum, L. wombeyi
Mniarogekko: New Caledonian Mossy Geckos; 2 species; New Caledonia
  M. chahoua (VU), *M. jalu*
Mokopirirakau: New Zealand Forest Geckos, 4 species; New Zealand
  M. cryptozoicus*, M. granulatus*, M. kahutarae*, M. nebulosus*
Naultinus: New Zealand Green Geckos; 8 species; New Zealand
  N. elegans*, N. gemmeus*, N. grayii*, N. manukanus*, N. punctatus*, N. rudis*,
  N. stellatus*, N. tuberculatus*
Nebulifera: Robust Velvet Gecko; 1 species; central eastern Australia
  N. robusta
Oedodera: New Caledonian Thick-necked Gecko; 1 species; New Caledonia
  O. marmorata (CR)
Oedura: Velvet Geckos, 9 species; Australia
  O. castelnaui, O. coggeri, O. filicipoda, O. gemmata, O. gracilis, O. jowalbinna,
  O. marmorata, O. monilis, O. tryoni
Paniegekko: Panié Massif Forest Gecko; 1 species; New Caledonia
  P. madjo
Pseudothecadactylus: Australian Giant Geckos; 3 species; northern Australia
  P. australis, P. cavaticus, P. lindneri
Rhacodactylus: New Caledonian Giant Geckos; 4 species; New Caledonia
  R. auriculatus, R. leachianus, R. trachycephalus, R. trachyrhynchus (EN)
Rhynchoedura: Beaked Geckos; 6 species; Australia
  R. angusta, R. eyrensis, R. mentalis, R. ormsbyi, R. ornata, R. sexapora
Strophurus: Spiny-tailed Geckos; 17 species; Australia
  S. assimilis, S. ciliaris, S. elderi, S. intermedius, S. jeanae, S. krisalys, S. mcmillani,
  S. michaelseni, S. rankini, S. robinsoni, S. spinigerus, S. strophurus, S. taeniatus,
  S. taenicauda, S. wellingtonae, S. williamsi, S. wilsoni
Toropuku: New Zealand Striped Gecko; 1 species; Stephens Island, Maud Island,
  New Zealand
  T. stephensi* (VU)
Tukutuku: Harlequin Gecko; 1 species; Stewart Island, New Zealand
  T. rakiurae*
Woodworthia: New Zealand Brown Geckos, 3 species; New Zealand
  W. brunnea*, W. chrysosireticus*, W. maculata*

Family Eublepharidae

Aeluroscalabotes: Cat Gecko; 1 species; Southeast Asia, Indonesia
   A. felinus

Coleonyx: Banded Geckos; 7 species; southwestern United States, Central America
   C. brevis, C. elegans, C. fasciatus, C. mitratus, C. reticulatus, C. switaki, C. variegatus

Eublepharis: Leopard Geckos; 5 species; western and southern Asia
   E. angramainyu, E. fuscus, E. hardwickii, E. macularius, E. turcmenicus

Goniurosaurus: Cave Geckos; 13 species; China, Vietnam, Ryukyu Islands (Japan), Hainan Island (China)
   G. araneus, G. bawanglingensis, G. catbaensis, G. hainanensis, G. huuliensis, G. kuroiwae (EN), *G. lichtenfelderi, G. luii, G. orientalis, G. splendens, G. toyamai, G. yamashinae, G. yingdeensis*

Hemitheconyx: Fat-tailed Geckos; 2 species; West Africa, northeast Africa
   H. caudicinctus, H. taylori

Holodactylus: Somali Clawed Geckos; 2 species; eastern Africa
   H. africanus, H. cornii

Family Gekkonidae

Afroedura: Flat Geckos; 15 species; southern Africa
   A. africana, A. amatolica, A. bogerti, A. halli, A. hawequensis, A. karroica, A. langi, A. loveridgei, A. major, A. marleyi, A. multiporis, A. nivaria, A. pondolia, A. tembulica, A. transvaalica

Afrogecko: African Leaf-toed Geckos; 4 species; southern Africa
   A. ansorgii, A. plumicaudus, A. porphyreus, A. swartbergensis

Agamura: Spider Gecko; 1 species; Iran, Afghanistan, Pakistan
   A. persica

Ailuronyx: Bronze Geckos; 3 species; Seychelles
   A. seychellensis, A. tachyscopaeus, A. trachygaster (VU)

Alsophylax: Straight-fingered Geckos; 6 species; Iran, Afghanistan, Central Asia, China, Mongolia
   A. laevis, A. loricatus, A. pipiens, A. przewalskii, A. szczerbaki, A. tadjikiensis

Altiphylax: Himalayan Geckos; 5 species (allocation of some species to this genus is uncertain); Afghanistan, India (Kashmir), Pakistan, Kyrgyzstan
   A. baturensis, A. levitoni, A. stoliczkai, A. tokobajevi, A. yarkandensis

Blaesodactylus: Madagascan Velvet Geckos; 4 species; Madagascar
   B. ambonihazo, B. antongilensis, B. boivini (VU), *B. sakalava*

Bunopus: Rock Geckos; 4 species; Middle East to Pakistan, Arabian Peninsula
   B. blanfordii, B. crassicauda, B. spatalurus, B. tuberculatus

Calodactylodes: Golden Geckos; 2 species; south India, Sri Lanka
   C. aureus, C. illingworthorum

Chondrodactylus: Giant Ground and Giant Thick-toed Geckos; 6 species; southern and eastern Africa
   C. angulifer, C. bibronii, C. fitzsimonsi, C. laevigatus, C. pulitzerae, C. turneri

Christinus: Australian Leaf-toed Geckos; 3 species; southern Australia, Norfolk Island, Lord Howe Island Group
   C. alexanderi, C. guentheri (VU), *C. marmoratus*

Cnemaspis: Forest Day Geckos; 103 species; equatorial Africa, India, Sri Lanka, Southeast Asia, Indian Ocean islands, western Indonesia
   C. affinis, C. africana, C. alantika, C. alwisi, C. amith, C. anaikattiensis (CR),

C. andersonii, C. argus, C. assamensis, C. aurantiacopes, C. australis, C. barbouri,
C. baueri, C. bayuensis, C. beddomei, C. biocellata, C. boiei, C. boulengerii,
C. caudanivea, C. chanardi, C. chanthaburiensis, C. clivicola, C. dezwaani,
C. dickersoni, C. dilepis, C. dringi, C. elgonensis, C. flavigaster, C. flavolineata,
C. gemunu, C. gigas, C. goaensis, C. gracilis, C. harimau, C. heteropholis,
C. huaseesom, C. indica, C. indraneildasii, C. jacobsoni, C. jerdonii, C. kallima,
C. kamolnorranathi, C. kandiana, C. karsticola, C. kendallii, C. koehleri,
C. kolhapurensis, C. kumarasinghei, C. kumpoli, C. laoensis, C. latha, C. limi,
C. littoralis, C. mcguirei, C. menikay, C. modiglianii, C. molligodai, C. monachorum,
C. monticola, C. mysoriensis, C. nairi, C. narathiwatensis, C. neangthyi, C. nigridius,
C. nilagirica, C. niyomwanae, C. nuicamensis, C. occidentalis, C. ornata, C. otai,
C. paripari, C. pava, C. pemanggilensis, C. perhentianensis, C. petrodroma,
C. phillipsi, C. phuketensis, C. podihuna, C. pseudomcguirei, C. psychedelica, C. pulchra,
C. punctata, C. punctatonuchalis, C. quattuorseriata, C. retigalensis, C. roticanai,
C. samanalensis, C. scalpensis, C. shahruli, C. siamensis, C. silvula, C. sisparensis,
C. spinicollis, C. timoriensis (probably not a member of this genus), *C. tropidogaster,*
C. tucdupensis, C. upendrai, C. uzungwae, C. vandeventeri, C. whittenorum, C. wicksi,
C. wynadensis, C. yercaudensis

Colopus: Southern African Ground Geckos; 2 species; Namibia, Botswana, South Africa, Zimbabwe, Zambia, Angola
   C. kochii, C. wahlbergii

Crossobamon: Comb-toed Geckos; 2 species; northwest India, Pakistan, Central Asia
   C. eversmanni, C. orientalis

Cryptactites: Salt Marsh Gecko; 1 species; South Africa
   C. peringueyi

Cyrtodactylus: Bent-toed Geckos; 171 species (allocation of some species to this genus is uncertain); Pakistan and Tibet to southeast Asia, Indonesia, Philippines, New Guinea, northern Australia, Solomon Islands
   C. aaroni, C. adleri, C. adorus, C. aequalis, C. agamensis, C. agusanensis,
   C. angularis, C. annandalei, C. annulatus, C. arcanus, C. aurensis, C. auribalteatus,
   C. australotitiwangsaensii, C. ayeyarwadyensis, C. badenensis, C. baluensis, C. batik,
   C. battalensis, C. batucolus, C. bichnganae, C. bidoupimontis, C. bintangredah,
   C. bintangtinggi, C. biordinis, C. boreoclivus, C. brevidactylus, C. brevipalmatus,
   C. buchardi, C. bugiamapensis, C. caovansungi, C. capreoloides, C. cattienensis,
   C. cavernicolus (VU), *C. chanhomeae, C. chauquangensis, C. chrysopylos, C. condorensis,*
   C. consobrinoides, C. consobrinus, C. cracens, C. cryptus, C. cucphuongensis,
   C. darmandvillei, C. dattanensis, C. derongo, C. deveti, C. dumnuii, C. durio,
   C. edwardtaylori, C. eisenmanae, C. elok, C. epiroticus, C. erythrops, C. fasciolatus,
   C. feae, C. fraenatus, C. fumosus, C. gansi, C. gordongekkoi, C. grismeri, C. gubaot,
   C. gubernatoris, C. halmahericus, C. himalayanus, C. hontreensis, C. hoskini,
   C. huongsonensis, C. huynhi, C. ingeri, C. interdigitalis, C. intermedius,
   C. irianjayaensis, C. irregularis, C. jambangan, C. jarakensis, C. jarujini, C. jellesmae,
   C. khasiensis, C. kimberleyensis, C. klugei, C. laevigatus, C. langkiawensis, C. lateralis,
   C. lawderanus, C. leegrismeri, C. lekaguli, C. lomyenensis, C. loriae, C. louisiadensis,
   C. macrotuberculatus, C. majulah, C. malayanus, C. malcolmsmithi, C. mamanwa,
   C. mandalayensis, S. markuscombaii, C. marmoratus, C. martini, S. martinstolli,
   C. matsuii, C. mcdonaldi, C. medioclivus, C. medogense, C. mimikanus, C. mintoni,
   C. murua, C. nepalensis, C. nigriocularis, C. novaeguineae, C. nuaulu, C. oldhami,

C. pageli, C. pantiensis, C. papilionoides, C. papuensis, C. paradoxus, C. payacola,
C. peguensis, C. philippinicus, C. phongnhakebangensis, C. phuquocensis, C. phuketensis,
C. pronarus, C. pseudoquadrivirgatus, C. pubisulcus, C. pulchellus, C. quadrivirgatus,
C. ramboda, C. redimiculus, C. robustus, C. roesleri, C. rubidus, C. russelli, C. sadleiri,
C. salomonensis, C. semenanjungensis, C. seribuatensis, C. sermowaiensis, C. serratus,
C. slowinskii, C. soba, C. spinosus, C. stresemanni, C. subsolanus, C. sumonthai,
C. sumuroi, C. surin, C. sworderi, C. takouensis, C. tamaiensis, C. tautbatorum,
C. teyniei, C. thirakhupti, C. thochuensis, C. tibetanus, C. tigroides, C. tiomanensis,
C. trialatofasciatus, C. tripartitus, C. tuberculatus, C. variegatus, C. wakeorum,
C. wallacei, C. wayakonei, C. wetariensis, C. yangbayensis, C. yoshii, C. zhaoermii,
C. ziegleri, C. zugi

Cyrtopodion: Palearctic Naked-toed Geckos; 23 species (allocation of some species to this genus is uncertain); Middle East, Arabian Peninsula, southwest Asia to northern India; introduced in Armenia, Red Sea coast of Africa
C. agamuroides, C. aravallense, C. baigii, C. belaense, C. brevipes, C. dadunense,
C. fortmunroi, C. gastrophole, C. golubevi, C. indusoani, C. kachhense, C. kiabii,
C. kirmanense, C. kohsulaimanai, C. mansarulum, C. montiumsalsorum,
C. persepolense, C. potoharense, C. rhodocaudum, C. rohtasfortai, C. scabrum,
C. sistanense, C. watsoni

Dixonius: Southeast Asian Leaf-toed Geckos; 5 species; Southeast Asia
D. aaronbaueri, D. hangseesom, D. melanostictus, D. siamensis, D. vietnamensis

Ebenavia: Madagascan Clawless Geckos; 2 species; Madagascar, Comores, Pemba Island (Tanzania)
E. inunguis, E. maintimainty (EN)

Elasmodactylus: Weak-skinned Geckos; 2 species; southeast Africa
E. tetensis, E. tuberculosus

Geckoella —Indian Ground Geckos; 7 species; India, Sri Lanka
G. albofasciatus, G. collegalensis, G. deccanensis, G. jeyporensis, G. nebulosus,
G. triedrus, G. yakhuna

Geckolepis: Fish-scale Geckos; 3 species; Madagascar, Comores
G. maculata, G. polylepis, G. typica

Gehyra: Dtellas or Stump-toed Geckos; 38 species; southeast Asia, Indonesia, Philippines, Ryukyu Islands (Japan), Australia, Pacific islands; introduced in Madagascar, Mascarene Islands, Seychelles, United States, and Mexico
G. angusticaudata, G. australis, G. baliola, G. barea (EN), *G. borroloola,*
G. brevipalmata, G. butleri, G. catenata, G. dubia, G. fehlmanni, G. fenestra,
G. georgpotthasti, G. insulensis, G. intermedia, G. interstitialis, G. kimberleyi, G. koira,
G. lacerata, G. lampei, G. lazelli, G. leopoldi, G. marginata, G. membranacruralis,
G. minuta, G. montium, G. mutilata, G. nana, G. occidentalis, G. oceanica, G. pamela,
G. papuana, G. pilbara, G. punctata, G. purpurascens, G. robusta, G. variegata,
G. vorax, G. xenopus

Gekko: Typical Geckos; 49 species; Tropical Asia, Japan, Taiwan, Indonesia, New Guinea, Solomon Islands, Vanuatu
G. albofasciolatus, G. athymus, G. auriverrucosus, G. badenii, G. canaensis,
G. canhi, G. carusadensis, G. chinensis, G. coi, G. crombota, G. ernstkelleri (VU),
G. gecko, G. gigante (VU), *G. grossmanni, G. hokouensis, G. japonicus, G. kikuchii,*
G. lauhachindai, G. melli, G. mindorensis, G. monarchus, G. nutaphandi,
G. palawanensis, G. palmatus, G. petricolus, G. porosus, G. reevesii, G. romblon,

G. rossi, G. russelltraini, G. scabridus, G. scientiadventura, G. shibatai, G. siamensis,
G. similignum, G. smithii, G. subpalmatus, G. swinhonis (VU), *G. taibaiensis,*
G. takouensis, G. tawaensis, G. taylori, G. truongi, G. verreauxi, G. vertebralis,
G. vietnamensis, G. vittatus, G. wenxianensis, G. yakuensis

Goggia: African Dwarf Leaf-toed Geckos; 8 species; South Africa, Namibia
G. braacki, G. essexi, G. gemmula, G. hewitti, G. hexapora, G. lineata,
G. microlepidota, G. rupicola

Hemidactylus: Split-toed Geckos; 123 species; northern South America, tropical
Africa, Mediterranean, Middle East, Arabian Peninsula, tropical and subtropical
Asia, Indonesia, New Guinea, Pacific islands (probably introduced); introduced
in the United States, Mexico, West Indies, parts of South and Central America,
northern Australia
H. aaronbaueri, H. agrius, H. albofasciatus, H. albopunctatus, H. alkiyumii,
H. anamallensis, H. angulatus, H. ansorgii, H. aporus, H. aquilonius, H. arnoldi,
H. barbierii, H. barodanus, H. bavazzanoi, H. bayonii, H. beninensis, H. boavistensis,
H. bouvieri, H. bowringii, H. brasilianus, H. brookii, H. citernii, H. craspedotus,
H. curlei, H. dawudazraqi, H. depressus, H. dracaenacolus (CR)*, H. echinus,*
H. endophis, H. fasciatus, H. festivus, H. flaviviridis, H. forbesii, H. foudaii,
H. frenatus, H. funaiolii, H. garnotii, H. giganteus, H. gleadowi, H. gracilis,
H. granchii, H. graniticolus, H. granti, H. greefii, H. gujaratensis, H. haitianus,
H. hajarensis, H. homoeolepis, H. hunae, H. imbricatus, H. inexpectatus,
H. inintellectus, H. isolepis, H. ituriensis, H. jubensis, H. jumailae, H. kamdemtohami,
H. karenorum, H. klauberi, H. kushmorensis, H. laevis, H. lamaensis, H. lankae,
H. laticaudatus, H. lavadeserticus, H. lemurinus, H. leschenaultii, H. longicephalus,
H. lopezjuradoi, H. luqueorum, H. mabouia, H. macropholis, H. maculatus,
H. makolowodei, H. masirahensis, H. matschiei, H. megalops, H. mercatorius,
H. mindiae, H. modestus, H. muriceus, H. newtoni, H. ophiolepis, H. ophiolepoides,
H. oxyrhinus, H. palaichthus, H. parvimaculatus, H. paucituberculatus, H. persicus,
H. pieresii, H. platycephalus, H. platyurus, H. prashadi, H. principensis,
H. pseudomuriceus, H. puccionii, H. pumilio, H. reticulatus, H. richardsonii,
H. robustus, H. romeshkanicus, H. ruspolii, H. saba, H. sataraensis, H. scabriceps,
H. shihraensis, H. sinaitus, H. smithi, H. somalicus, H. squamulatus, H. stejnegeri,
H. subtriedrus, H. tanganicus, H. tasmani, H. taylori, H. tenkatei, H. thayene,
H. treutleri, H. triedrus, H. tropidolepis, H. turcicus, H. vietnamensis, H. yerburyi

Hemiphyllodactylus: Gypsy Geckos; 8 species; southern China, southeast Asia,
Taiwan, Philippines, Indonesia, Melanesia, Micronesia, Polynesia
H. aurantiacus, H. ganoklonis, H. harterti, H. insularis, H. margarethae,
H. titiwangsaensis, H. typus, H. yunnanensis

Heteronotia: Prickly Geckos; 3 species; Australia
H. binoei, H. planiceps, H. spelea

Homopholis: African Velvet Geckos; 4 species; eastern and southern Africa
H. arnoldi, H. fasciata, H. mulleri, H. wahlbergii

Lepidodactylus: Scaly-toed Geckos; 33 species; southeast Asia, China, Taiwan,
Ryukyu and Bonin Islands (Japan), Philippines, Indonesia, New Guinea, Pacific
islands; introduced in Australia, Hawaii, parts of tropical America
L. aureliit, L. aureolineatus, L. balioburius, L. browni, L. buleli, L. christiani,
L. euaensis, L. flaviocularis, L. gardineri, L. guppyi, L. herrei, L. intermedius,
L. listeri (EN)*, L. lombocensis, L. lugubris, L. magnus, L. manni, L. moestus,*
L. mutahi, L. novaeguineae, L. oligoporus, L. oortii, L. orientalis, L. paurolepis,

L. planicaudus, L. pulcher, L. pumilus, L. ranauensis, L. shebae, L. tepukapili,
L. vanuatuensis, L. woodfordi, L. yami

Luperosaurus: Fringed Geckos; 13 species; Philippines, Borneo, Sumatra, peninsular Malaysia

L. angliit, L. brooksii, L. browni, L. corfieldi, L. cumingii, L. gulat, L. iskandari,
L. joloensis (EN), *L. kubli, L. macgregori* (EN), *L. palawanensis, L. sorok, L. yasumai*

Lygodactylus: Dwarf Day Geckos; 62 species; Madagascar, sub-Saharan Africa, central South America

L. angolensis, L. angularis, L. arnoulti, L. bernardi, L. bivittis (VU), *L. blancae,*
L. blanci (VU), *L. bradfieldi, L. broadleyi, L. capensis, L. chobiensis, L. conradti,*
L. conraui, L. decaryi, L. depressus, L. expectatus, L. fischeri, L. grandisonae,
L. graniticolus, L. gravis (VU), *L. guibei, L. gutturalis, L. heterurus, L. howelli,*
L. incognitus, L. inexpectatus, L. insularis, L. intermedius (EN), *L. keniensis,*
L. kimhowelli, L. klemmeri, L. klugei, L. lawrencei, L. madagascariensis (VU),
L. manni, L. methueni (VU), *L. miops, L. mirabilis* (CR), *L. mombasicus,*
L. montanus, L. montiscaeruli, L. nigropunctatus, L. ocellatus, L. ornatus (EN),
L. pauliani, L. picturatus, L. pictus, L. rarus, L. rex, L. roavolana (EN), *L. scheffleri,*
L. scorteccii, L. somalicus, L. soutpansbergensis, L. stevensoni, L. thomensis,
L. tolampyae, L. tuberosus, L. verticillatus, L. waterbergensis, L. wetzeli, L. williamsi

Matoatoa: Ghost Geckos; 2 species; Madagascar

M. brevipes (VU), *M. spannringi* (CR)

Mediodactylus: Palearctic Naked-toed Geckos; 14 species (allocation of some species to this genus is uncertain); southern Italy, Balkan Peninsula, Turkey, Middle East, Central Asia, Iran, Pakistan

M. amictopholis (EN), *M. aspratilis, M. brachykolon, M. dehakroensis, M. heterocercus,*
M. heteropholis, M. ilamensis, M. kotschyi, M. narynensis, M. russowii, M. sagittifer,
M. spinicaudus, M. stevenandersoni, M. walli

Microgecko: Iranian Dwarf Geckos; 4 species; Iraq, Iran, Pakistan

M. depressus, M. helenae, M. latifi, M. persicus

Nactus: Indo-Pacific Naked-toed Geckos; 12 species; northeastern Australia, New Guinea, Indonesia, Pacific islands, islands off Mauritius, Réunion (extinct)

N. acutus, N. cheverti, N. coindemirensis (VU), *N. eboracensis, N. galgajuga,*
*N. kunan, N. multicarinatus, N. pelagicus, N. serpensinsula** (VU), *N. soniae†,*
N. sphaerodactylodes, N. vankampeni

Narudasia: Festive Gecko; 1 species; Namibia

N. festiva

Pachydactylus: Thick-toed Geckos; 57 species; southern and east central Africa

P. acuminatus, P. affinis, P. amoenus, P. atorquatus, P. austeni, P. barnardi, P. bicolor,
P. boehmei, P. capensis, P. caraculicus, P. carinatus, P. etultra, P. fasciatus, P. formosus,
P. gaiasensis, P. geitje, P. goodi, P. griffini, P. haackei, P. katanganus, P. kladaroderma,
P. kobosensis, P. labialis, P. laevigatus, P. latirostris, P. macrolepis, P. maculatus,
P. maraisi, P. mariquensis, P. mclachlani, P. monicae, P. montanus, P. monticolus,
P. namaquensis, P. oculatus, P. oreophilus, P. oshaughnessyi, P. otaviensis, P. parascutatus,
P. punctatus, P. purcelli, P. rangei, P. reconditus, P. robertsi, P. rugosus, P. sansteynae,
P. scherzi, P. scutatus, P. serval, P. tigrinus, P. tsodiloensis, P. vansoni, P. vanzyli,
P. visseri, P. waterbergensis, P. weberi, P. werneri

Paragehyra: Paragehyras, 2 species; Madagascar

P. gabriellae (EN), *P. petiti* (VU)

Paroedura: Madagascan Ground Geckos; 16 species; Madagascar, Comores
  P. androyensis (VU), *P. bastardi, P. gracilis, P. homalorhina, P. ibityensis, P. karstophila,*
  P. lohatsara (CR), *P. maingoka, P. masobe* (EN), *P. oviceps, P. picta, P. sanctijohannis*
  (EN), *P. stumpffi, P. tanjaka* (EN), *P. vahiny, P. vazimba* (VU)
Perochirus: Micronesian Geckos; 3 species; Micronesia, Vanuatu, Guam (extirpated)
  P. ateles, P. guentheri, P. scutellatus
Phelsuma: Day Geckos; 51 species; Madagascar, southeastern Africa and islands off
  the east coast, Andaman and Nicobar Islands (India); introduced in United States
  P. abbotti*, P. andamanense*, P. antanosy* (CR), *P. astriata*, P. barbouri*, P. berghofi*,*
  P. borai*, P. borbonica*, P. breviceps* (VU), *P. cepediana*, P. comorensis*,*
  P. dorsivittata*, P. dubia*, P. edwardnewtonii*†, P. flavigularis* (EN), *P. gigas*†,*
  P. gouldi*, P. grandis*, P. guentheri* (EN), *P. guimbeaui*, P. guttata*, P. hielscheri**
  (VU), *P. hoeschi*, P. inexpectata*, P. kely*, P. klemmeri** (EN), *P. kochi*, P. laticauda*,*
  P. lineate*, P. madagascariensis*, P. malamakibo*, P. masohoala* (CR), *P. modesta*,*
  P. mutabilis*, P. nigristriata* (VU), *P. ornata*, P. parkeri*, P. parva*, P. pronki** (CR),
  P. pusilla*, P. quadriocellata*, P. ravenala*, P. robertmertensi* (EN), *P. roesleri** (EN),
  P. rosagularis*, P. seippi* (EN), *P. serraticauda** (EN), *P. standingi** (VU),
  P. sundbergi*, P. vanheygeni* (EN), *P. v-nigra**
Pseudoceramodactylus: Gulf Sand Gecko; 1 species; eastern Arabian Peninsula
  P. khobarensis
Pseudogekko: False Geckos; 4 species; Philippines
  P. brevipes (VU), *P. compressicorpus, P. labialis, P. smaragdinus*
Ptenopus: Barking Geckos; 3 species; southern Africa
  P. carpi, P. garrulus, P. kochi
Ptychozoon: Parachute Geckos; 8 species; northeast India, Burma (Myanmar),
  Southeast Asia, western Indonesia, Philippines
  P. horsfieldii, P. intermedium, P. kaengkrachanense, P. kuhli, P. lionotum,
  P. nicobarensis, P. rhacophorus, P. trinotaterra
Rhinogecko: Snouted Spider Geckos; 2 species; Iran, Pakistan
  A. femoralis, A. misonnei
Rhoptropella: Namaqua Day Gecko; 1 species; southwestern Africa
  R. ocellata
Rhoptropus: Namib Day Geckos; 7 species; Angola, Namibia
  R. afer, R. barnardi, R. biporosus, R. boultoni, R. bradfieldi, R. diporus, R. taeniostictus
Stenodactylus: Short-fingered Geckos; 12 species; Africa, Middle East, Arabian
  Peninsula
  S. affinis, S. arabicus, S. doriae, S. grandiceps, S. leptocosymbotus, S. mauritanicus,
  S. petrii, S. pulcher, S. slevini, S. stenurus, S. sthenodactylus, S. yemenensis
Tenuidactylus: Central Asian Naked-toed Geckos; 6 species (allocation of some
  species to this genus is uncertain); Caucasus, Iran, Afghanistan, Central Asia,
  western China, Southern Mongolia
  T. caspius, T. elongatus, T. fedtschenkoi, T. longipes, T. turcmenicus, T. voraginosus
Tropiocolotes: Dwarf Sand Geckos; 9 species; northern Africa, Arabian Peninsula,
  Israel, Jordan
  T. algericus, T. bisharicus, T. nattereri, T. nubicus, T. scortecci, T. somalicus, T. steudneri,
  T. tripolitanus, T. wolfgangboehmei
Urocotyledon: Prehensile-Tailed Geckos; 5 species; equatorial Africa, Seychelles
  U. inexpectata, U. palmata, U. rasmusseni, U. weileri, U. wolterstorffi

Uroplatus: Madagascan Leaf-Tailed Geckos; 14 species; Madagascar
  U. alluaudi* (VU), *U. ebenaui** (VU), *U. fimbriatus**, *U. finiavana**, *U. giganteus** (VU),
  U. guentheri* (EN), *U. henkeli** (VU), *U. lineatus**, *U. malahelo** (EN), *U. malama**
  (VU), *U. phantasticus**, *U. pietschmanni** (EN), *U. sameiti**, *U. sikorae**

Family Phyllodactylidae
Asaccus: Southwest Asian Leaf-toed Geckos; 16 species; Turkey, Arabian Peninsula
  A. andersoni, A. barani, A. caudivolvulus, A. elisae, A. gallagheri, A. granularis,
  A. griseonotus, A. iranicus, A. kermanshahensis, A. kurdistanensis, A. montanus,
  A. nasrullahi, A. platyrhynchus, A. saffinae, A. tangestanensis, A. zagrosicus
Garthia: Chilean Marked Geckos; 2 species; southern South America
  G. gaudichaudii, G. penai
Gymnodactylus: Neotropical Naked-toed Geckos; 5 species; Brazil, Trinidad (?)
  G. amarali, G. darwinii, G. geckoides, G. guttulatus, G. vanzolinii
Haemodracon: Socotran Leaf-toed Geckos; 2 species; Socotra (Arabian Sea)
  H. riebeckii, H. trachyrhinus
Homonota: Marked Geckos; 9 species; eastern South America
  H. andicola, H. borellii, H. darwinii, H. fasciata, H. rupicola, H. underwoodi,
  H. uruguayensis, H. whitii, H. williamsii
Phyllodactylus: American Leaf-toed Geckos; 52 species; Western Americas from
  California to Chile, Galápagos Islands (Ecuador), northwestern South America,
  Puerto Rico, Hispaniola, Barbados, Curaçao, Aruba, Bonaire
P. angelensis, P. angustidigitus, P. apricus, P. barringtonensis, P. baurii, P. bordai,
  P. bugastrolepis, P. clinatus, P. darwini, P. davisi, P. delcampoi, P. delsolari, P. dixoni,
  P. duellmani, P. galapagensis, P. gerrhopygus, P. gilberti, P. heterurus, P. hispaniolae,
  P. homolepidurus, P. inaequalis, P. insularis, P. interandinus, P. johnwrighti, P. julieni,
  P. kofordi, P. lanei, P. leei (VU)*, P. lepidopygus, P. martini, P. microphyllus, P. muralis,*
  P. nocticolus, P. palmeus, P. papenfussi, P. partidus, P. paucituberculatus, P. pulcher,
  P. pumilius, P. reissii, P. rutteni, P. santacruzensis, P. sentosus, P. sommeri, P. thompsoni,
  P. tinklei, P. transversalis, P. tuberculosus, P. unctus, P. ventralis, P. wirshingi, P. xanti
Phyllopezus: Brazilian Geckos; 4 species; central South America east of the Andes
  P. lutzae, P. maranjonensis, P. periosus, P. pollicaris
Ptyodactylus: Fan-footed Geckos; 6 species; North Africa Sahel, Middle East,
  Arabian Peninsula, Pakistan
  P. guttatus, P. hasselquistii, P. homolepis, P. oudrii, P. puiseuxi, P. ragazzii
Tarentola: Wall Geckos; 29 species; northern and central Africa, Mediterranean
  Europe, Atlantic islands, West Indies, Jamaica (now extinct); introduced United
  States, Uruguay
  T. albertschwartzi†, T. americana, T. angustimentalis, T. annularis, T. bischoffi,
  T. boavistensis, T. bocagei, T. boehmei, T. boettgeri, T. caboverdiana, T. chazaliae,
  T. crombiei, T. darwini, T. delalandii, T. deserti, T. ephippiata, T. fogoensis, T. gigas,
  T. gomerensis, T. maioensis, T. mauritanica, T. mindiae, T. neglecta, T. nicolauensis,
  T. parvicarinata, T. protogigas, T. raziana, T. rudis, T. substituta
Thecadactylus: Turnip-tailed Geckos; 3 species; Mexico, Central America, northern
  South America (Amazon drainage and north), West Indies
  T. oskrobapreinorum, T. rapicauda, T. solimoensis

Family Pygopodidae

Aprasia: Australian Worm Lizards; 12 species; southern and western Australia
   A. aurita (CR), *A. fusca, A. haroldi, A. inaurita, A. parapulchella, A. picturata,*
   A. pseudopulchella, A. pulchella, A. repens, A. rostrata (VU), *A. smithi, A. striolata*

Delma: Delmas, 19 species; Australia
   D. australis, D. borea, D. butleri, D. concinna, D. desmosa, D. elegans, D. fraseri,
   D. grayii, D. impar (VU), *D. inornata, D. labialis* (VU), *D. mitella, D. molleri,*
   D. nasuta, D. pax, D. plebeia, D. tealei, D. tincta, D. torquata (VU)

Lialis: Snake Lizards; 2 species; Australia, New Guinea
   L. burtonis, L. jicari

Ophidiocephalus: Bronzeback Snake Lizard; 1 species; central Australia
   O. taeniatus (VU)

Paradelma: Brigalow Scaly-foot; 1 species; central east Australia
   P. orientalis (VU)

Pletholax: Slender Slider; 1 species; coastal Western Australia
   P. gracilis

Pygopus: Scaly-foots; 5 species; Australia
   P. lepidopodus, P. nigriceps, P. robertsi, P. schraderi, P. steelescotti

Family Sphaerodactylidae

Aristelliger: Croaking Geckos; 8 species; Mexico, Belize, islands of Western
   Caribbean, Atlantic islands of Colombia
   A. barbouri, A. cochranae, A. expectatus, A. georgeensis, A. hechti, A. lar, A. praesignis,
   A. reyesi

Chatogekko: Pug-nosed Gecko; 1 species; northern South America
   C. amazonicus

Coleodactylus: Pygmy Geckos; 5 species; northern South America
   C. brachystoma, C. elizae, C. meridionalis, C. natalensis, C. septentrionalis

Euleptes: European Leaf-toed Gecko; 1 species; Tyrrhenian Sea, central
   Mediterranean
   E. europaea

Gonatodes: American Day Geckos; 29 species; northern South America, Central
   America, West Indies; introduced United States, Galápagos Islands (Ecuador)
   G. albogularis, G. alexandermendesi, G. annularis, G. antillensis, G. astralis,
   G. atricucullaris, G. caudiscutatus, G. ceciliae, G. concinnatus, G. daudini (CR),
   G. eladioi, G. falconensis, G. hasemani, G. humeralis, G. infernalis, G. lichenosus,
   G. ligiae, G. nascimentoi, G. ocellatus, G. petersi, G. purpurogularis, G. riveroi,
   G. rozei, G. seigliei, G. superciliaris, G. taniae, G. tapajonicus, G. timidis, G. vittatus

Lepidoblepharis: Scaly-eyed Geckos; 18 species; southern Central America, parts of
   northern South America
   L. buchwaldi, L. colombianus, L. conolepis, L. duolepis, L. festae, L. grandis,
   L. heyerorum, L. hoogmoedi, L. intermedius, L. microlepis, L. miyatai,
   L. montecanoensis, L. oxycephalus, L. peraccae, L. ruthveni, L. sanctaemartae,
   L. williamsi, L. xanthostigma

Pristurus: Semaphore Geckos; 26 species; Arabian Peninsula, Iran, Pakistan, Socotra
   (Arabian Sea), northeastern Africa, Mauritania
   P. abdelkuri, P. adrarensis, P. carteri, P. celerrimus, P. collaris, P. crucifer,
   P. flavipunctatus, P. gallagheri, P. gasperetti, P. guichardi, P. insignis, P. insignoides,

P. longipes, P. mazbah, P. minimus, P. obsti, P. ornithocephalus, P. phillipsii, P. popovi, P. rupestris, P. saada, P. samhaensis, P. schneideri, P. simonettai, P. sokotranus, P. somalicus

Pseudogonatodes: Clawed Geckos; 7 species; northwestern South America
   P. barbouri, P. furvus, P. gasconi, P. guianensis, P. lunulatus, P. manessi, P. peruvianus

Quedenfeldtia: Moroccan Day Geckos, 2 species; Morocco
   Q. moerens, Q. trachyblepharus

Saurodactylus: Lizard-fingered Geckos; 3 species; northwest Africa
   S. brosseti, S. fasciatus (VU), *S. mauritanicus*

Sphaerodactylus: Dwarf Geckos; 99 species; south Florida, West Indies, southern Mexico to northern South America
   S. altavelensis, S. argivus, S. argus, S. ariasae, S. armasi (EN), *S. armstrongi, S. asterulus, S. beattyi, S. becki, S. bromeliarum, S. caicosensis, S. callocricus* (VU), *S. celicara, S. cinereus, S. clenchi, S. cochranae, S. continentalis, S. copei, S. corticola, S. cricoderus, S. cryphius, S. darlingtoni, S. difficilis, S. dimorphicus, S. docimus, S. dunni, S. elasmorhynchus, S. elegans, S. elegantulus, S. epiurus, S. fantasticus, S. gaigeae, S. gilvitorques, S. glaucus, S. goniorhynchus, S. graptolaemus, S. guanajae, S. heliconiae, S. homolepis, S. inaguae, S. intermedius, S. kirbyi* (VU), *S. klauberi, S. ladae, S. lazelli, S. leonardoraldesi, S. leucaster, S. levinsi, S. lineolatus, S. macrolepis, S. mariguanae, S. microlepis, S. micropithecus* (EN), *S. millepunctatus, S. molei, S. monensis, S. nicholsi, S. nigropunctatus, S. notatus, S. nycteropus, S. ocoae, S. oliveri, S. omoglaux, S. oxyrhinus, S. pacificus, S. parkeri, S. parthenopion, S. parvus, S. perissodactylius, S. phyzacinus, S. pimiento* (EN), *S. plummeri, S. ramsdeni, S. randi, S. rhabdotus, S. richardi, S. richardsonii, S. roosevelti, S. rosaurae, S. ruibali, S. sabanus, S. samanensis, S. savagei, S. scaber, S. scapularis* (VU), *S. schuberti, S. schwartzi, S. semasiops, S. shrevei, S. siboney, S. sommeri, S. sputator, S. storeyae* (EN), *S. streptophorus, S. thompsoni, S. torrei* (VU), *S. townsendi, S. underwoodi, S. vincenti, S. williamsi* (CR), *S. zygaena*

Teratoscincus: Wonder Geckos or Frog-eyed Geckos; 7 species; Iran, Pakistan, Arabian Peninsula, Central Asia to western China
   T. bedriagai, T. keyserlingii, T. microlepis, T. przewalskii, T. roborowskii, T. scincus, T. toksunicus

Appendix B

Organizations and Publications Devoted (in Part) to the Study of Geckos

Herpetological societies and institutions whose missions include the promotion of the study or conservation of reptiles, including geckos. Publications of these groups are indicated in italics and the country in which each is based is placed in parentheses.

American Society of Ichthyologists and Herpetologists, *Copeia* (USA)

Amphibia and Reptiles Research Organization of Sri Lanka, *Lyriocephalus* (Sri Lanka)

Asociación Herpetologica Argentina, *Cauadernos de Herpetologia* (Argentina)

Asociación Herpetologica Española, *Basic and Applied Herpetology* (formerly *Revista Española de Herpetología*), *Boletin de la Asociación Herpetologica Española* (Spain)

Australasian Affiliation of Herpetological Societies, *Herpetofauna* (Australia)

Brazilian Society of Herpetology, *South American Journal of Herpetology* (Brazil)

The British Herpetological Society, *The Herpetological Bulletin*, *The Herpetological Journal* (United Kingdom)

Chengdu Institute of Biology, *Asian Herpetological Research* (formerly *Asiatic Herpetological Research*) (China)

Chicago Herpetological Society, *Bulletin of the Chicago Herpetological Society* (USA)

Chinese Herpetological Society, *Herpetologica Sinica* (China)

Deutsche Gesellschaft für Herpetologie und Terrarienkunde, *Elaphe*, *Salamandra* (Germany)

Global Gecko Association, *Chit Chat*, *Gekko* (International)

Herpetological Association of Africa, *African Herp News*, *African Journal of Herpetology* (South Africa)

Herpetological Society of Japan, *Bulletin of the Herpetological Society of Japan*, *Current Herpetology* (Japan)

The Herpetologists' League, *Herpetologica*, *Herpetological Monographs* (USA)

International Reptile Conservation Foundation, *IRCF Reptiles &
Amphibians* (online only from 2012; USA)
Israel Herpetological Information Center, *Hardun* (Israel)
Madras Crocodile Bank Trust, *Hamadryad* (India)
Madras Snake Park, *Cobra* (India)
Maryland Herpetological Society, *Bulletin of the Maryland Herpetological
Society* (USA)
Nederlandse Vereniging voor Herpetologie en Terrariumkunde, *Lacerta*
(The Netherlands)
Nikolsky Herpetological Society, *Russian Journal of Herpetology* (Russia)
Nordisk Herpetologisk Forening, *Nordisk Herpetologisk Forening-bladet*
(Denmark)
Österreichischen Gesellschaft für Herpetologie, *Herpetozoa* (Austria)
Societas Europaea Herpetologica, *Amphibia-Reptilia, Herpetology Notes*
(pan-European)
Societas Herpetologica Italica, *Acta Herpetologica* (Italy)
Société Herpetologique de France, *Bulletin de la Société Herpetologique de
France* (France)
Society for the Study of Amphibians and Reptiles, *Herpetological Review,
Journal of Herpetology* (USA)
Suomen Herpetologinen Yhdistys Ry, *Herpetomania* (Finland)
Sveriges Herpetologiska Riksförening, *Snoken* (Sweden)
Terrariengemeinschadt Berlin e.V., *Sauria* (Germany)
World Congress of Herpetology (Global)

Other herpetological or general zoological journals publishing papers
about geckos (many additional journals, including museum journals and
many national, regional, and local journals, at least occasionally publish
papers on geckos; the titles listed here are representative only). Those
marked with an asterisk are no longer being published.

Acta Anatomica
Acta Zoologica
African Zoology
American Museum Novitates
Amphibian and Reptile Conservation (online only)
*Applied Herpetology
Australian Journal of Zoology
Biological Journal of the Linnean Society
Bonn Zoological Bulletin (formerly *Bonner Zoologische Beiträge*)
Breviora
Bulletin of the American Museum of Natural History

Bulletin of the Museum of Comparative Zoology
Canadian Journal of Zoology
Check List
*Dactylus
Fieldiana: Zoology
Gekkota
*herpetofauna (Germany not be confused with Herpetofauna from
   Australia, see above)
Herpetological Conservation and Biology (online only)
Herpetotropicos
Integrative and Comparative Biology
Israel Journal of Zoology
Journal of Experimental Biology
Journal of Morphology
Journal of Natural History
Journal of the Bombay Natural History Society
Journal of the German Egyptian Society of Zoology
Journal of Threatened Taxa
Journal of Zoology, London
Molecular Phylogenetics and Evolution
New Zealand Journal of Zoology
North-West Journal of Zoology
Phyllomedusa
Physiological and Biochemical Zoology
Podarcis (online only)
Proceedings of the Academy of Natural Sciences of Philadelphia
*Proceedings of the Biological Society of Washington
Proceedings of the California Academy of Sciences
Raffles Bulletin of Zoology
Records of the Australian Museum
Records of the Western Australian Museum
Reptiles
Reptilia (Spanish, English, and German editions)
*Scientific Publications of the Natural History Museum of the University of
   Kansas*
Sichuan Journal of Zoology
Taprobanica
Tropical Zoology
Turkish Journal of Zoology
Vertebrate Zoology
*The Vivarium
Zoo Keys

Zoologica Scripta
Zoological Journal of the Linnean Society
Zoological Science
Zoological Systematics and Evolutionary Research
Zoologischer Anzeiger
Zoology
Zoology in the Middle East
Zootaxa

Alifanov, V. R. 1990. The oldest gecko (Lacertilia, Gekkonidae) from the lower Cretaceous of Mongolia. *Paleontological Journal* 23 (1): 128–131.

Anderson, S. C., and A. E. Leviton. 1966. A new species of *Eublepharis* from Southwestern Iran (Reptilia: Gekkonidae). *Occasional Papers of the California Academy of Sciences* (53): 1–5.

Arnold, E. N. 1990. The two species of Moroccan day-geckoes, *Quedenfeldtia* (Reptilia: Gekkonidae). *Journal of Natural History* 24: 757–762.

Arnold, E. N. 1993. Historical changes in the ecology and behaviour of semaphore geckos (*Pristutrus*, Gekkonidae) and their relatives. *Journal of Zoology, London* 229: 353–384.

Arnold, E. N., and G. Poinar. 2008. A 100 million year old gecko with sophisticated adhesive toe pads, preserved in amber from Myanmar. *Zootaxa* 1847: 62–68.

Autumn K., A. Dittmore, D. Santos, M. Spenko, and M. Cutkosky. 2006. Frictional adhesion: A new angle on gecko attachment. *Journal of Experimental Zoology* 209: 3569–3579.

Autumn, K., D. Jindrich, D. F. DeNardo, and R. Mueller. 1999. Locomotor performance at low temperature and the evolution of nocturnality in lizards. *Evolution* 53: 580–599.

Autumn K., M. Sitti, Y. A. Liang, A. M. Peattie, W. R. Hansen, S. Sponberg, T. W. Kenny, R. Fearing, J. N. Israelachvili, and R. J. Full. 2002. Evidence for van der Waals adhesion in gecko setae. *Proceedings of the National Academy of Sciences of the United States* 99: 12252–12256.

Bannock, C. A., A. H. Whitaker, and G. J. Hickling. 1999. Extreme longevity of the common gecko (*Hoplodactylus maculatus*) on Motunau Island, Canterbury, New Zealand. *New Zealand Journal of Ecology* 23: 101–103.

Bartlett, M. D., Croll, A. B., King, D. R., Irschick, D. J., and Crosby, A. J. 2012. Looking beyond fibrillar features to scale gecko-like adhesion. *Advanced Materials* 24: 1078–1083.

Bauer, A. M. 1989. Extracranial endolymphatic sacs in *Eurydactylodes* (Reptilia: Gekkonidae), with comments on endolymphatic function in lizards in general. *Journal of Herpetology* 23: 172–175.

Bauer, A. M. 1990. Gekkonid lizards as prey of invertebrates and predators of vertebrates. *Herpetological Review* 21: 83–87.

Bauer, A. M. 2007. The foraging biology of the Gekkota: Life in the middle. In *Lizard Ecology: The Evolutionary Consequences of Foraging Mode in Lizards*, edited by S. M. Reilly, L. B. McBrayer, and D. B. Miles, pp. 371–404. Cambridge: Cambridge University Press.

Bauer, A. M., W. Böhme, and W. Weitschat. 2005. An Early Eocene gecko from Baltic amber and its implications for the evolution of gecko adhesion. *Journal of Zoology, London* 265: 327–332.

Bauer, A. M., and I. Das. 2000. Review of the gekkonid genus *Calodactylodes* (Reptilia: Squamata) from India and Sri Lanka. *Journal of South Asian Natural History* 5: 25–35.

Bauer, A. M., and A. P. Russell. 1986. *Hoplodactylus delcourti* n. sp. (Reptilia: Gekkonidae), the largest known gecko. *New Zealand Journal of Zoology* 13: 141–148.

Bauer, A. M., and A. P. Russell. 1994. Is autotomy frequency reduced in geckos with "actively functional" tails? *Herpetological Natural History* 2: 1–15.

Bauer, A. M., A. P. Russell, and R. E. Shadwick. 1989. Mechanical properties and morphological correlates of fragile skin in gekkonid lizards. *Journal of Experimental Biology* 145: 79–102.

Bauer, A. M., and R. A. Sadlier. 1994. Diet of the New Caledonian gecko *Rhacodactylus auriculatus* (Squamata, Gekkonidae). *Russian Journal of Herpetology* 1: 108–113.

Böhme, W. 1988. Zur Genitalmorphologie der Sauria: funktionelle und stammesgeschichtliche Aspekte. *Bonner Zoologische Monographien* 27: 1–176.

Böhme, W., and M. Sering. 1997. Tail squirting in *Eurydactylodes*: independent evolution of caudal defensive glands in a diplodactyline gecko (Reptilia, Gekkonidae). *Zoologischer Anzeiger* 235: 225–229.

Bontius, J. 1658. *Historiae Naturalis et Medicae Indiae Orientalis Libri Sex*. In *De Indiae Utriusque re Naturali et Medica*, by Willem Piso. Amstelaedami [Amsterdam]: Apud Ludovicum et Danielem Elzevirios.

Borsuk-Białynicka, M. 1990. *Gobekko cretacicus* gen. et sp. n.: A new gekkonid lizard from the Cretaceous of the Gobi Desert. *Acta Palaeontologica Polonic* 35: 67–76, pls. 17–18.

Brown, W. C., and A. C. Alcala. 1957. Viability of lizard eggs exposed to sea water. *Copeia* 1957: 39–41.

Burghardt, G. M. 2002. Play in reptiles. In *New Encyclopedia of Reptiles and Amphibians*, edited by T. R. Halliday, and K. Adler, pp. 112–113. Oxford: Oxford University Press.

Bustard, H. R. 1967. Gekkonid lizards adapt fat storage to desert environments. *Science* 158: 1197–1198.

Bustard, H. R. 1968. The ecology of the Australian gecko, *Gehyra variegata*, in northern New South Wales. *Journal of Zoology, London* 154: 113–138.

Bustard, H. R. 1969. The population ecology of the Australian Geckos *Diplodactylus williamsi* and *Gehyra australis* in northern New South Wales. *Proceedings of the Koninklijke Nederlandse Akademie Van Wetenschappen* 72: 1–32.

Bustard, H. R. 1971. A population study of the eyed gecko, *Oedura ocellata* Boulenger, in northern New South Wales, Australia. *Copeia* 1971: 658–669.

Chou, L. M., C. F. Leong, and B. L. Choo. 1988. The role of optic, auditory and olfactory senses in prey hunting by two species of geckos. *Journal of Herpetology* 22: 349–351.

Cogger, H. G. 2000. *Reptiles and Amphibians of Australia*, 6th ed. Sanibel Island, FL: Ralph Curtis Books.

Congdon, J. D., L. J. Vitt, and W. W. King. 1974. Geckos: adaptive significance and energetics of tail autotomy. *Science* 184: 1379–1380.

Conrad, J. L. 2008. Phylogeny and systematic of Squamata (Reptilia) based on morphology. *Bulletin of the American Museum of Natural History* 310: 1–182.

Cooper, W. E., Jr. 1995. Prey chemical discrimination and foraging mode in gekkonoid lizards. *Herpetological Monographs* 9: 120–129.

Cree, A., 1994. Low annual reproductive output in female reptiles from New Zealand. *New Zealand Journal of Zoology* 21: 351–372.

Cuellar, O., and A. G. Kluge. 1972. Natural parthenogenesis in the gekkonid lizard *Lepidodactylus lugubris*. *Journal of Genetics* 61: 14–26.

Cunkelman, A. A. 2005. The ecology of *Rhacodactylus leachianus* and *R trachyrhynchus* in an area of sympatry. Unpublished M. S. thesis, Villanova University.

de Vosjoli, P., F. Fast, and A. Repashy. 2003. *Rhacodactylus, the Complete Guide to their Selelction and Care.* Vista, CA: Advanced Visions.

de Vosjoli, P., R. Tremper, and R. Klingenberg. 2005. *The Herpetoculture of Leopard Geckos: Twenty-seven Generations of Living Art.* Vista, CA: Advanced Visions.

de Silva, A., I. Das, A. M. Bauer, and S. Goonawardene. 2004. Vedda rock art in Sri Lanka depicting reptiles, with special reference to golden gecko *Calodactylodes illingworthorum* (Reptilia: Gekkonidae). *Lyriocephalus* 5: 213–219, pl. 8.

Dial, B. E., and L. C. Fitzpatrick. 1981. The energetic costs of tail autotomy to reproduction in the lizard *Coleonyx brevis* (Sauria: Gekkonidae). *Oecologia* 51: 310–317.

Dial, B. E., and K. Schwenk. 1996. Olfaction and predator detection in *Coleonyx brevis* (Squamata: Eublepharidae), with comments on the functional significance of buccal pulsing in geckos. *Journal of Experimental Zoology* 276: 415–424.

Dunson, W. A. 1982. Low water vapor conductance of hard-shelled eggs of the gecko lizards *Hemidactylus* and *Lepidodactylus*. *Journal of Experimental Zoology* 219: 377–379.

Estes, R. 1983. *Sauria Terrestria, Amphisbaenia. Handbuch der Paläoherpetologie.* Part 10A. Stuttgart, Germany: Gustav Fischer Verlag.

FitzSimons, V. F. 1943. *The Lizards of South Africa. Memoirs of the Transvaal Museum.*

Forbes, P. 2006. *The Gecko's Foot: Bio-Inspiration: Engineering New Materials from Nature.* New York: Norton.

Frankenberg, E. 1978. Interspecific and seasonal variation of daily activity times in gekkonid lizards (Reptilia, Lacertilia). *Journal of Herpetology* 12: 505–519.

Gamble, T., A. M. Bauer, G. R. Colli, E. Greenbaum, T. R. Jackman, L. J. Vitt, and A. M. Simons. 2011. Coming to America: Multiple origins of New World geckos. *Journal of Evolutionary Biology* 24: 231–244.

Gamble, T., A. M. Bauer, E. Greenbaum, and T. R. Jackman. 2008a. Evidence for Gondwanan vicariance in an ancient clade of gecko lizards. *Journal of Biogeography* 35: 88–104.

Gamble, T., A. M. Bauer, E. Greenbaum, and T. Jackman. 2008b. Out of the blue: cryptic higher level taxa and a novel, trans-Atlantic clade of gecko lizards (Gekkota, Squamata). *Zoologica Scripta* 37: 355–366.

Gamble, T., E. Greenbaum, A. P. Russell, T. R. Jackman, and A. M. Bauer. 2012. Repeated origin and loss of toepads in geckos. *PLoS ONE* 7 (6): e39429.

Gans, C., and P. F. A. Maderson. 1973. Sound producing mechanisms in recent reptiles: review and comment. *American Zoologist* 13: 1195–1203.

Gibbons, J. R. H., and F. Clunie. 1984. Brief notes on the voracious gecko *Gehyra vorax*. *Domodomo* 2: 34–36.

Girling, J. E., A. Cree, and L. J. Guillette, Jr. 1998. Oviducal structure in four species of gekkonid lizard differing in parity mode and eggshell structure. *Reproduction, Fertility and Development* 10: 139–154.

Greer, A. E. 1989. *The Biology and Evolution of Australian Lizards.* Chipping Norton, NSW, Australia: Surrey Beatty.

Grismer, L. L. 1988. Phylogeny, taxonomy, classification and biogeography of eublepharid geckos. In *Phylogenetic Relationships of the Lizard Families*, edited by R. Estes, and G. Pregill, 369–469. Stanford, CA: Stanford University Press.

Haacke, W. D. 1975. The burrowing geckos of southern Africa, 1 (Reptilia: Gekkonidae). *Annals of the Transvaal Museum* 29: 197–243, plates 10–11.

Hansen, D. M., and C. B. Müller. 2009. Reproductive ecology of the endangered enigmatic Mauritian endemic *Roussea* simplex (Rousseaceae). *International Journal of Plant Sciences* 170: 42–52.

Hansen, W. R., and K. Autumn. 2010. Evidence for self-cleaning in gecko setae. *Proceedings of the National Academy of Sciences of the United States* 102: 385–389.

Hedges, S. B., and R. Thomas. 2001. At the lower size limit in amniote vertebrates: A new diminutive lizard from the West Indies. *Caribbean Journal of Science* 37: 168–173.

Henkel, F. W., and W. Schmidt. 1995. *Geckoes: Biology, Husbandry and Reproduction.* Malabar, FL: Krieger Publishing Company.

Henkel, F. W., and W. Schmidt. 2003. *Professional Breeders Series: Geckos.* Frankfurt-am-Main, Germany: Edition Chimaira.

Henle, K. 1990. Population ecology and life history of three terrestrial geckos in arid Australia. *Copeia* 1990: 759–781.

Herrel, A., F. De Vree, V. Delheusy, and C. Gans. 1999. Cranial kinesis in gekkonid lizards. *Journal of Experimental Biology* 202: 3687–3698.

Hibbitts, T. J., M. J. Whiting, and D. M. Stuart-Fox. 2007. Shouting the odds: Vocalization signals status in a lizard. *Behavioral Ecology and Sociobiology* 61: 1169–1176.

Huey, R. B., E. R. Pianka, and L. J. Vitt. 2001. How often do lizards "run on empty"? *Ecology* 82: 1–7.

Huey, R. B., J. J. Tewksbury, C. A. Deutsch, L. J. Vitt, P. E. Hertz, and H. J. Alvarez Perez. 2009. Why tropical forest lizards are vulnerable to climate warming. *Proceedings of the Royal Society B: Biological Sciences* 276: 1939–1948.

Ibargüengoytia, N. R., and L. M. Casalins. 2007. Reproductive biology of the southernmost gecko *Homonota darwini*: Convergent life-history patterns among Southern Hemisphere reptiles living in harsh environments. *Journal of Herpetology* 41: 72–80.

Irschick, D. J., C. C. Austin, K. Petren, R. N. Fisher, J. B. Losos, and O. Ellers. 1996. A comparative analysis of clinging ability among pad-bearing lizards. *Biological Journal of Linnean Society* 59: 21–35.

Jamniczky, H. A., A. P. Russell, M. K. Johnson, S. J. Montuelle, and V. L. Bels. 2009. Morphology and histology of the tongue and oral chamber of *Eublepharis macularius* (Squamata: Gekkonidae), with special reference to the foretongue and its role in fluid uptake and transport. *Evolutionary Biology* 36: 397–406.

Jennings, W. B., E. R. Pianka, and S. Donnellan. 2003. Systematics of the lizard family Pygopodidae with implications for the diversification of Australian temperate biotas. *Systematic Biology* 52: 757–780.

Kearney, M., R. Shine, S. Comber, and D. Pearson. 2001. Why do geckos group? An analysis of "social" aggregations in two species of Australian lizards. *Herpetologica* 57: 411–422.

Kluge, A. G. 1967. Higher taxonomic categories of gekkonid lizards and their evolution. *Bulletin of the American Museum of Natural History* 135: 1–60, pls. 1–5.

Kluge, A. G. 1976. Phylogenetic relationships in the lizard family Pygopodidae: An evaluation of theory, methods and data. *Miscellaneous Publications of the Museum of Zoology, University of Michigan* (152): [i–iv], 1–72.

Kluge, A. G. 1987. Cladistic relationships in the Gekkonoidea (Squamata, Sauria). *Miscellaneous Publications of the Museum of Zoology, University of Michigan* (173): i–iv, 1–54.

Kluge, A. G. 1995. Cladistic relationships of sphaerodactyl lizards. *American Museum Novitates* (3139): 1–23.

Kluge, A. G. 2001. Gekkotan lizard taxonomy. *Hamadryad* 26: 1–209.

Kluge, A. G., and M. J. Eckardt. 1969. *Hemidactylus garnotii* Duméril and Bibron, a tripoid all-female species of gekkonid lizards. *Copeia* 1969: 651–664.

Köhler, G. 2005. *Incubation of Reptile Eggs*. Malabar, FL: Krieger Publishing.

Kratochvíl, L., and D. Frynta. 2005. Egg shape and size allometry in geckos (Squamata: Gekkota), lizards with contrasting eggshell structure: Why lay spherical eggs? *Journal of Zoological Science* 44: 217–222.

Kratochvíl, L., and L. Kubicka. 2007. Why reduce clutch size to one or two eggs? Reproductive allometries reveal different evolutionary causes of invariant clutch size in lizards. *Functional Ecology* 21: 171–177.

Kraus, F. 2009. *Alien Reptiles and Amphibians, A Scientific Compendium, and Analysis*. Dordrecht, The Netherlands: Springer Verlag.

Lamb, T, and A. M. Bauer. 2006. Footprints in the sand: Independent reduction of subdigital lamellae in the Namib-Kalahari burrowing geckos. *Proceedings of the Royal Society B: Biological Sciences* 273: 855–864.

Laurenti, J. N. 1768. *Specimen Medicum, Exhibens Synopsin Reptilium Emendatum cum Experimentis circa Venena et Antidota Reptilium Austriacorum, quod Authoritate et Consensu*. Viennæ [Vienna]: Joan. Thomae nob. De Trattnern.

Lee, M. S. Y., P. M. Oliver, and M. N. Hutchinson. 2008. Phylogenetic uncertainty and molecular clock calibrations: A case study of legless lizards (Pygopodidae, Gekkota). *Molecular Phylogenetics and Evolution* 50: 661–666.

Linnaeus, C. 1758. *Systema Naturæ per Regna Tria Naturæ, Secundum Classes, Ordines, Genera, Species, cum Characteribus, Differentiis, Synonymis, Locis*. Edito decima. Tomus I. Holmiae [Stockholm]: Laurentii Salvii.

Loveridge, A. 1947. Revision of the African lizards of the family Gekkonidae. *Bulletin of the Museum of Comparative Zoology* 98: 1–469.

Mader, D. R. 2006. *Reptile Medicine and Surgery*, 2nd ed. St. Louis, MO: Saunders.

Maderson, P. F. A. 1970. Lizard glands and lizard hands: models for evolutionary study. *Forma et Functio* 3: 179–204.

Mahdavi A., et al. 2008. A biodegradable and biocompatible gecko-inspired tissue adhesive. *Proceedings of the National Academy of Sciences of the United States* 105: 2307–2312.

Manley, G. A., and J. E. M. Kraus. 2010. Exceptional high-frequency hearing and matched vocalizations in Australian pygopod geckos. *Journal of Experimental Biology* 213: 1876–1885.

Marcellini, D. 1977. The function of a vocal display of the lizard *Hemidactylus frenatus* (Sauria: Gekkonidae). *Animal Behaviour* 25: 414–417.

Mitchell, J. C. 1986. Cannibalism in reptiles: A worldwide review. *SSAR Herpetological Circular* 15.

Moore, B. A., A. P. Russell, and A. M. Bauer. 1991. The structure of the larynx of the tokay gecko (*Gekko gecko*), with particular reference to the vocal cords and glottal lips. *Journal of Morphology* 210: 227–238.

Moritz, C. 1987. Parthenogenesis in the tropical gekkonid lizard, *Nactus arnouxii* (Sauria: Gekkonidae). *Evolution* 41: 1252–1266.

Olesen, J. M., and A. Valido. 2003. Lizards as pollinators and seed dispersers: An island phenomenon. *Trends in Ecology and Evolution* 18: 177–180.

Oliver, P. M., and K. L. Sanders. 2009. Molecular evidence for Gondwanan origins of multiple lineages within a diverse Australasian gecko radiation. *Journal of Biogeography* 36: 2044–2055.

Patchell, F. C., and R. Shine. 1986. Food habits and reproductive biology of the Australian legless lizards (Pygopodidae). *Copeia* 1986: 30–39.

Peattie, A. M. 2008. Subdigital setae of narrow-toed geckos, including a eublepharid (*Aeluroscalabotes felinus*). *Anatomical Record* 291: 869–875.

Petren, K., D. T. Bolger, and T. J. Case. 1993. Mechanisms in the competitive success of an invading sexual gecko over an asexual native. *Science* 259: 354–358.

Pianka, E. R. 1986. *Ecology and Natural History of Desert Lizards*. Princeton, NJ: Princeton University Press.

Pianka, E. R., and R. B. Huey. 1978. Comparative ecology, niche segregation, and resource utilization among gekkonid lizards in the southern Kalahari. *Copeia* 1978: 691–701.

Pianka, E. R., and H. D. Pianka. 1976. Comparative ecology of twelve species of nocturnal lizards (Gekkonidae) in the Western Australian desert. *Copeia* 1976: 125–142.

Pianka, E. R., and S. S. Sweet. 2005. Integrative biology of sticky feet in geckos. *Bioessays* 27: 647–652.

Pianka, E. R., and L. J. Vitt. 2003. *Lizards: Windows to the Evolution of Diversity*. Berkeley, CA: University of California Press.

Piantoni, C., N. R. Ibargüengoytia, and V. E. Cussac. 2006. Growth and age of the southernmost distributed gecko of the world (*Homonota darwini*) studied by skeletochronology. *Amphibia-Reptilia* 27: 393–400.

Pike, D. A., R. M. Andrews, and W.-G. Du. 2012. Eggshell morphology and gekkotan life-history evolution. *Evolutionary Ecology* 26: 847–861.

Regalado, R. 2003. Roles of visual, acoustic, and chemical signals in social interactions of the tropical house gecko (*Hemidactylus mabouia*). *Caribbean Journal of Science* 39: 307–320.

Rehorek, S. J., B. T. Firth, and M. N. Hutchinson. 2000. The structure of the nasal chemosensory system in squamate reptiles: I. The olfactory organ, with special reference to olfaction in geckos. *Journal of Biosciences* 25: 173–179.

Rodda, G. H., G. Perry, R. J. Rondeau, and J. Lazell. 2001. The densest terrestrial vertebrate. *Journal of Tropical Ecology* 17: 331–338.

Rödder, D., M. Solé, and W. Böhme. 2008. Predicting the potential distributions of two alien invasive house geckos (Gekkonidae: *Hemidactylus frenatus*, *Hemidactylus mabouia*). *North-Western Journal of Zoology* 4: 236–246.

Röll, B. 2000. Gecko vision-visual cells, evolution, and ecological constraints. *Journal of Neurocytology* 29: 471–484.

Rosenberg, H. I., and A. P. Russell. 1980. Structural and functional aspects of tail squirting: A unique defense mechanism of *Diplodactylus* (Reptilia: Gekkonidae). *Canadian Journal of Zoology* 58: 865–881.

Rösler, H. 1995. *Geckos der Welt—Alle Gattungen*. Leipzig, Germany: Urania Verlag.

Rösler, H. 2000. Kommentierte Liste der rezent, subrezent und fossil bekannten Gecko-Taxa (Reptilia: Gekkonomorpha). *Gekkota* 2: 28–153.

Rösler, H. 2005. *Vermehrung von Geckos*. Offenbach, Germany: Herpeton, Verlag Elke Köhler.

Roth, L. S. V., and A. Kelber. 2004. Nocturnal colour vision in geckos. *Proceedings of the Royal Society B: Biological Sciences* (Supplement). 271: S485–S487.

Roth, L. S. V., L. Lundström, A. Kelber, R. H. H. Kröger, and P. Unsbo. 2009. The pupils and optical systems of gecko eyes. *Journal of Vision* 9 (3): 1–11.

Russell, A. P. 1975. A contribution to the functional analysis of the foot of the Tokay, *Gekko gecko* (Reptilia: Gekkonidae*). Journal of Zoology, London* 176: 437–476.

Russell, A. P. 1976. Some comments concerning interrelationships amongst gekkonine geckos. In *Morphology and Biology of Reptiles*, edited by A. d'A. Bellairs, and C. B. Cox, pp. 217–244. London: Academic Press.

Russell, A. P. 1979. Parallelism and integrated design in the foot structure of gekkonine and diplodactyline geckos. *Copeia* 1979: 1–21.

Russell, A. P. 2002. Integrative functional morphology of the gekkotan adhesive system (Reptilia: Gekkota). *Integrative and Comparative Biology* 42: 1154–1163.

Russell, A. P., and A. M. Bauer. 1990. Substrate excavation in the Namibian webfooted gecko, *Palmatogecko rangei* Andersson 1908, and its ecological significance. *Tropical Zoology* 3: 197–207.

Russell, A. P., and T. E. Higham. 2009. A new angle on clinging in geckos: Incline, not substrate, triggers the deployment of the adhesive system. *Proceedings of the Royal Society B: Biological Sciences* 276: 3705–3709.

Russell, A. P., and M. K. Johnson. 2007. Real-world challenges to, and capabilities of, the gekkotan adhesive system: Contrasting the rough and the smooth. *Canadian Journal of Zoology* 85: 1228–1238.

Schwenk, K. 1985. Occurrence, distribution and functional significance of taste buds in lizards. *Copeia* 1985: 91–101.

Schwenk, K. 1993. Are geckos olfactory specialists? *Journal of Zoology, London* 229: 289–302.

Seipp, R., and F. W. Henkel. 2011. *Rhacodactylus: Biology, Natural History & Husbandry*, 2nded. Frankfurt am Main, Germany: Edition Chimaira.

Szczerbak, N. N., and M. L. Golubev. 1996. *Gecko Fauna of the USSR and Contiguous Regions*. Ithaca, NY: Society for the Study of Amphibians and Reptiles.

Thompson, M. B., C. H. Daugherty, A. Cree, D. C. French, J. C. Gillingham, and R. E. Barwick. 1992. Status and longevity of the tuatara, *Sphenodon guntheri*, and Duvaucel's gecko, *Hoplodactylus duvaucelii*, on North Brother Island, New Zealand. *Journal of Royal Society of New Zealand* 22: 123–130.

Uetz, P. (compiler). 2011. The Reptile Database. http://www. reptile-database. org. Accessed 4 July 2012.

Underwood, G. 1954. On the classification and evolution of geckos. *Proceedings of the Zoological Society of London* 124: 469–492.

Underwood, G. 1957. On lizards of the family Pygopodidae: A contribution to the morphology and phylogeny of the Squamata. *Journal of Morphology* 100: 207–268.

Vidal, N., and S. B. Hedges. 2009. The molecular evolutionary tree of lizards, snakes, and amphisbaenians. *Comptes Rendus Biologies* 332: 129–139.

Viets, B. E., A. Tousignant, M. A. Ewert, C. E. Nelson, and D. Crews. 1993. Temperature-dependent sex determination in the leopard gecko, *Eublepharis macularius. Journal of Experimental Zoology* 265: 679–683.

Vinson, J., and J. M. Vinson. 1969. The saurian fauna of the Mascarene Islands. *The Mauritius Institute Bulletin* 6: 203–320.

Vitt, L. J. 1986. Reproductive tactics of sympatric gekkonid lizards with a comment on the evolutionary and ecological consequences of invariant clutch size. *Copeia* 1986: 773–786.

Vitt, L. J., E. R. Pianka, W. E. Cooper, and K. Schwenk. 2003. History and the global ecology of squamate reptiles. *American Naturalist* 162: 44–60.

Vitt, L. J., S. S. Sartorius, T. C. S. Avila-Pires, P. A. Zani, and M. C. Espósito. 2005. Small in a big world: Ecology of leaf-litter geckos in New World tropical forests. *Herpetological Monographs* 19: 137–152.

Vitt, L. J., and P. A. Zani. 1997. Ecology of the nocturnal lizard *Thecadactylus rapicauda* (Sauria: Gekkonidae) in the Amazon Region. *Herpetologica* 53: 165–179.

Webb, J. K., and R. Shine. 1994. Feeding habits and reproductive biology of Australian pygopodid lizards of the genus *Aprasia*. *Copeia* 1994: 390–398.

Werner, Y. L. 1972. Observations on eggs of eublepharid lizards, with comments on the evolution of the Gekkonoidea. *Zoologische Mededelingen* 47: 211–224, pl. 1.

Werner, Y. L. 1977. Ecological comments on some gekkonid lizards of the Namib Desert, South West Africa. *Madoqua* 10: 157–169.

Werner, Y. L. 1980. Apparent homosexual behavior in an all-female population of a lizard, *Lepidodactylus lugubris*, and its probable interpretation. *Zeitschrift für Tierpsychologie* 54: 144–150.

Werner, Y. L., and M. Broza. 1969. Hypothetical function of elevated locomotory postures in geckos (Reptilia: Gekkonidae). *Israel Journal of Zoology* 18: 349–355.

Werner, Y. L., E. Frankenberg, M. Volokita, and R. Harari. 1993. Longevity of geckos (Reptilia: Lacertilia: Gekkonoidea) in captivity: An analytical review incorporating new data. *Israel Journal of Zoology* 39: 105–124.

Werner, Y. L., S. Okada, H. Ota, G. Perry, and S. Tokunaga. 1997. Varied and fluctuating foraging modes in nocturnal lizards of the family Gekkonidae. *Asiatic Herpetological Research* 7: 153–165.

Werner Y. L., and T. Seifan. 2006. Eye size in geckos: Asymmetry, allometry, sexual dimorphism, and behavioral correlates. *Journal of Morphology* 267: 1486–1500.

Werner, Y. L., and E. G. Wever. 1972. The function of the middle ear in lizards: *Gekko gecko* and *Eublepharis macularius* (Gekkonoidea). *Journal of Experimental Zoology* 179: 1–16.

Whiting, M. J., L. T. Reaney, and J. S. Keogh. 2007. Ecology of Wahlberg's velvet gecko, *Homopholis wahlbergii*, in southern Africa. *African Zoology* 42: 38–44.

Zhang, X., T. G. Wensel, and C. Yuan. 2006. Tokay gecko photoreceptors achieve rod-like physiology with cone-like proteins. *Photochemistry and Photobiology* 82: 1452–1460.

Index

Page numbers in italic indicate figures.

Index